T0298695

Resilient Hybrid Electronics for Extreme/Harsh Environments

The success of future innovative technology relies upon a community with a shared vision. Here, we present an overview of the latest technological progress in the field of printed electronics for use in harsh or extreme environments. Each chapter unlocks scientific and engineering discoveries that will undoubtedly lead to progression from proof of concept to device creation.

The main topics covered in this book include some of the most promising materials, methods, and the ability to integrate printed materials with commercial components to provide the basis for the next generation of electronics that are dubbed "survivable" in environments with high g-forces, corrosion, vibration, and large temperature fluctuations. A wide variety of materials are discussed that contribute to robust hybrid electronics, including printable conductive composite inks, ceramics and ceramic matrix composites, polymer-derived ceramics, thin metal films, elastomers, solders and epoxies, to name a few. Collectively, these materials and associated components are used to construct conductive traces, interconnects, antennas, pressure sensors, temperature sensors, power inducting devices, strain sensors and gauges, soft actuators, supercapacitors, piezo ionic elements, resistors, waveguides, filters, electrodes, batteries, various detectors, monitoring devices, transducers, and RF systems and graded dielectric, or graded index (GRIN) structures. New designs that incorporate the electronics as embedded materials into channels, slots and other methods to protect the electronics from the extreme elements of the operational environment are also envisioned to increase their survivability while remaining cognizant of the required frequency of replacement, reapplication and integration of power sources. Lastly, the ability of printer manufacturers, software providers and users to work together to build multi-axis, multi-material, and commercial-off-the-shelf (COTS) integration into user-friendly systems will be a great advancement for the field of printed electronics.

Therefore, the blueprint for manufacturing resilient hybrid electronics consists of novel designs that exploit the benefits of advances in additive manufacturing that are then efficiently paired with commercially available components to produce devices that exceed known constraints. As a primary example, metals can be deposited onto polymers in a variety of ways, including aerosol jetting, microdispensing, electroplating, sintering, vacuum deposition, supersonic beam cluster deposition, and plasma-based techniques, to name a few. Taking these scientific discoveries and creatively combining them into robotic, multi-material factories of the future could be one shared aim of the printed electronics community toward survivable device creation.

Resilient Hybrid Electronics for Extreme/ Harsh Environments

Edited by
Amanda Schrand, L.J. Holmes,
and Eric MacDonald

CRC Press
Taylor & Francis Group
Boca Raton London New York

CRC Press is an imprint of the
Taylor & Francis Group, an **informa** business

First edition published 2024
by CRC Press
6000 Broken Sound Parkway NW, Suite 300, Boca Raton, FL 33487-2742

and by CRC Press
4 Park Square, Milton Park, Abingdon, Oxon, OX14 4RN

CRC Press is an imprint of Taylor & Francis Group, LLC

© 2024 selection and editorial matter, [Amanda Schrand, L.J. Holmes and Eric MacDonald]; individual chapters, the contributors

ISBN: 9780367687649 (hbk)
ISBN: 9780367687656 (pbk)
ISBN: 9781003138945 (ebk)

DOI: 10.1201/9781003138945

Typeset in Times
by codeMantra

Contents

Acknowledgments ix
Editors x
Contributors xii

Chapter 1   Introduction to Printed Electronics 1

             John Hornick

             Reference 3

Chapter 2   Which Printed Conductive Inks and Interconnects
Survive High G and Thermal Cycling? 4

             Clayton Neff, Edwin Elston, and Amanda Schrand

             2.1   Introduction 4
             2.2   Development of Testing Protocols for Ink Adhesion
and Impact Resilience 6
                     2.2.1   Scratch Adhesion Test (SAT) Tool 7
                     2.2.2   High G Testing 11
                     2.2.3   Summary of Testing Protocols for Printed
Electronics 12
             2.3   Evaluation of Adhesion, RF Performance, and
Interconnects Subject to Harsh Environments 13
                     2.3.1   Micro-Dispensed Conductive Inks 13
                     2.3.2   Aerosol Jet Conductive Inks 18
                     2.3.3   Summary of Printed Electronics Materials
Evaluation 23
              2.4   Conclusions and Future Outlook 24
              Acknowledgments 25
              References 25

Chapter 3   Additively Manufactured Antennas for Aerospace
Harsh Environments 31

             Eduardo A. Rojas-Nastrucci

             3.1   Introduction 31
              3.2   Additive Manufacturing Processes for Antennas 32
             3.3   Advanced Antennas for High-Temperature and
Hypersonic Applications 34
                     3.3.1   The Hypersonic Antenna Environment 34
                     3.3.2   High-Temperature Antennas and Sensors 35

　　　3.3.3　　Design of Antennas for Conical and Cylindrical
　　　　　　　Geometries 37
　3.4　Additively Manufactured Antennas for Space Missions 37
　3.5　Conclusions 40
　References 40

Chapter 4　Printed Pressure Sensors for Extreme Environments 46

Tosin D. Ajayi, Spencer Nguyen, Amanda Schrand,
and Chengying Xu

　4.1　Introduction 46
　　　4.1.1　　Brief Introduction to Additive
　　　　　　　Manufacturing (AM) 47
　　　4.1.2　　Pressure Sensors 47
　4.2　Printing Techniques for Pressure Sensors
　　　Fabrication 48
　　　4.2.1　　Fused Deposition Modeling 48
　　　4.2.2　　Ultrasonic Additive Manufacturing (UAM) 51
　　　4.2.3　　Selective Laser Processes 53
　　　4.2.4　　Inkjet Printing (IJP) 56
　4.3　Other AM Methodologies for Fabricating
　　　Advanced Pressure Sensors 59
　4.4　Printed Sensors through Polymer-Derived
　　　Ceramic (PDC) Process 60
　　　4.4.1　　Printed Polymer-Derived Ceramics 60
　　　4.4.2　　Printed PDC Sensor 61
　4.5　Conclusions 64
　Acknowledgments 65
　References 65

Chapter 5　Metallization of 3D-Printed Devices 69

Nathan Lazarus

　5.1　Introduction 69
　5.2　Approaches for Metallization 70
　　　5.2.1　　Electroplating 70
　　　5.2.2　　Electroless Plating 72
　　　5.2.3　　Liquid Metals 74
　　　5.2.4　　Sintering 76
　　　5.2.5　　Sputtering and Evaporation 79
　5.3　Application Example: Inductors and Wireless Power 80
　5.4　Conclusion 84
　Acknowledgments 85
　References 85

Chapter 6   Printing Electrically Conductive Patterns on Polymeric and
3D-Printed Systems...93

Lorenzo Migliorini, Tommaso Santaniello, and Paolo Milani

6.1   Introduction 93
6.2   Conductive Traces on Transparent Flexible Substrates..........96
6.3   Conductive Traces Implanted in Elastomeric/Stretchable
Materials..96
6.4   Metallization of Ionogels for Actuation, Energy
Harvesting and Storage 100
6.5   Conductive Traces on 3D-Printed Objects: 3D-Printed
Electronics..104
6.6   Conclusion 108
References 108

Chapter 7   Direct Write Printed Electronics and Materials Synthesis Using
Non-equilibrium Plasma-Based Techniques 114

*Matthew Montgomery, Chris Funke, Neal Magdefrau,
Paul Sheedy, Mary Herndon, Wayde Schmidt,
Andrew S. Morgan, Evyatar Shaulsky,
and Shomeek Mukhopadhyay*

7.1   Introduction 115
7.2   Direct Write Plasma-Based Techniques 116
7.3   Background of the Technology... 118
7.4   Deposition of Bulk Metals...120
7.4.1   Copper 120
7.4.2   Zinc..123
7.5   Other Complex Structures..125
7.6   Conclusion 126
Acknowledgments..126
References 126

Chapter 8   Additively Manufactured Ceramics with Embedded
Conductors for High-Temperature Applications 129

*Eleanor Rogenski, Victoria Admas,
Bhargavi Mummareddy, Bradley Duncan,
Eric MacDonald, and Pedro Cortes*

8.1   Conclusions...139
Acknowledgments 140
References 140

Chapter 9   Considerations for Design and Manufacturing of Flex
                 Devices and Printed Conductive Elements 143

                 Ken Blecker

                 9.1   Introduction 143
                 9.2   Thermal Considerations 149
                 9.3   Manufacturing Process.. 152
                 9.4   Environment and Usage... 153
                 9.5   Lifecycle Consideration.. 156
                 9.6   Case Studies.. 157
                 9.7   Conclusion 158
                 Acknowledgments 159
                 References 159

Chapter 10 Three-Dimensional Functional RF Devices Enabled through
                 Additive Manufacturing 161

                 Austin Good, Zachary Larimore, and Paul Parsons

                 10.1   Introduction: Additive as the Most Viable
                          Manufacturing Method.. 161
                 10.2   Additive Provides Rapid Prototyping and
                          Extreme Customization of RF Devices............................... 163
                 References 165

Index.. 167

Acknowledgments

This book is the result of years of research and was conceived in the midst of building the foundations of printed electronics for harsh/extreme environments during the COVID-19 pandemic. It is no surprise that the contributing authors, editors and publisher have a strong track record of persevering and producing an excellent product. As a reflection, I can see the development of my own writing and publishing from peer-reviewed papers to book chapters and now as a book editor. For all of those efforts, there is a clear leader, someone who inspires and pushes the team forward toward sharing the impact and understanding of the research. This does not diminish the efforts of the team. In fact, the results of efficient team work can be exponentially greater in scope and depth than trying to accomplish everything alone. Partnerships provide true value in the research community, and the authors of this book were deliberately chosen to cover perspectives in Government, Academia, Industry, Small Businesses and Research Institutes including the Air Force Research Laboratory, United States Army Combat Capabilities Development Command, United States Army Research Laboratory, Embry-Riddle Aeronautical University, Florida State University, Lincoln Laboratory Massachusetts Institute of Technology, North Carolina State University, University of Texas at El Paso, Yale University, Youngstown State University, University of Milan-Italy, Electron Microscopy Innovative Technologies LLC, Mat-Chem LLC, Plazmod Technologies Inc, Raytheon Technologies Center, Samtec, Sciperio, Raytheon UMass Lowell Research Institute and the University of Dayton Research Institute. Even the editorial team has unique strengths that have made this work possible. I'd like to thank Eric MacDonald for sharing his strength in quickly providing relevant feedback as he has honed those skills of focus and depth to lead two journals to success. Similarly, I'd like to thank L.J. Holmes for providing valuable connections and information to carry us to the finish line. It is my hope that this book will serve as reference for those entering the field of printed electronics as well as a contemplation for the next generation of leaders to carry the vision forward.

A special thanks to Alexis Schrand for designing the cover art. Her creative talents were recognized early on, and may they continue to be nurtured.

"Your value will not be what you know; it will be what you share."—Ginni Rometty (former President & CEO of IBM, 2012–2020)

—**Editor-in-Chief, Amanda Schrand, Ph.D.**

Editors

Amanda Schrand, Ph.D., currently serves as a senior engineer and group leader for the development of resilient, hybrid additively manufactured electronics at the Munitions Directorate of the Air Force Research Laboratory (AFRL) at Eglin Air Force Base, Florida. She is the principal investigator for several Cross Service efforts on 3D-printed conformal antennas, frequency selective surfaces, precision electrodes for fiber waveguides, pressure/temperature sensors, strain gauges and high-voltage circuits, to name a few. Her efforts in ceramics printing innovation have resulted in two patents with commercialization and licensing of the technology. Dr. Schrand received her doctoral degree in Materials Science and Engineering from the University of Dayton with the dual support of the Dayton Area Graduate Studies Institute (DAGSI) and the Oak Ridge Associated Universities (ORAU) fellowships. She has fostered a multi-disciplinary career over the past 20-plus years to gain experience in a range of medical, science and engineering fields. Her written work has been published in many professional venues including *Nature Protocols*, and her article on Additive Manufacturing in *Defense* is required reading for the Air War College. She has been honored by many individual and team awards including the Team Eglin Women's History Month Trailblazer award recognizing her contributions to leadership and mentorship. She is an active member of the Institute of Electrical and Electronics Engineers (IEEE) professional society, and she recently began chairing the Women's Panel on Career Development in RF Technology in addition to an International forum on Women in Additive Manufacturing in Italy and proposed collaborative work with the UK.

L.J. Holmes Jr. is the executive director of Research and Engineering at Harrisburg University of Science and Technology, where he leads the development and operation of an Advanced Manufacturing Research Institute. The mission of this academic institute is to create an interdisciplinary forum for bringing materials, processing and manufacturing together by digital design and innovative manufacturing methods. Mr. Holmes left federal service in 2018 after 15 years at the US Army Research Laboratory (ARL). His final posting at ARL was the director of Research Partnerships and Communication for the ARL Center for Agile Materials Manufacturing Science (CAMMS). He was also the lead for ARL's Hybrid Manufacturing research portfolio, including the management of materials and manufacturing science programs related to multi-material processing technologies for functional/multi-functional devices. Mr. Holmes is also the director of Government Relations at nScrypt in Orlando, FL. nScrypt designs and manufactures high-precision micro-dispensing and direct digital manufacturing equipment with unmatched accuracy and flexibility. Mr. Holmes is also the chief of manufacturing at the Applied Science and Research Organization of America (ASTRO America). ASTRO America is a non-profit, non-partisan research institute and think tank dedicated to advancing public interest through manufacturing and technology.

Eric MacDonald, Ph.D., is professor of Aerospace and Mechanical Engineering and Murchison Chair at the University of Texas at El Paso, and he serves as the associate dean of Research and Graduate Studies for the College of Engineering. Dr. MacDonald received his doctoral degree (2002) in Electrical and Computer Engineering from the University of Texas at Austin. He worked in industry for 12 years at IBM and Motorola. Subsequently, he co-founded a start-up specializing in CAD software, and the startup was acquired by a firm in Silicon Valley. Dr. MacDonald held faculty fellowships at NASA's Jet Propulsion Laboratory, US Navy Research and was awarded a US State Department Fulbright Fellowship in South America. His research interests include 3D-printed multi-functional applications and process monitoring in additive manufacturing with instrumentation and computer vision for improved quality and yield. As a co-founding editor of the Elsevier journal *Additive Manufacturing*, MacDonald has helped direct the academic journal to have the highest impact factor among all manufacturing journals worldwide. He has recently been involved in the commissioning of a second partner journal, *Additive Manufacturing Letters*, upon which he serves as the editor-in-chief. His recent projects include 3D printing of structures such as nano satellites with structurally embedded electronics—one of which was launched into Low Earth Orbit in 2013 and a replica of which was on display at the London Museum of Science. He has over 100 peer-reviewed publications and dozens of patents, one of which was licensed by Sony and Toshiba from IBM. He is a member of ASME and ASEE, senior member of IEEE, and a registered Professional Engineer in the US state of Texas.

Contributors

Victoria Admas
Youngstown State University
Youngstown, Ohio, USA

Tosin D. Ajayi
Department of Mechanical and
  Aerospace Engineering
North Carolina State University
Raleigh, North Carolina, USA

Ken Blecker
United States Army Combat
  Capabilities Development
  Command
Picatinny Arsenal, New Jersey, USA

Pedro Cortes
Youngstown State University
Youngstown, Ohio, USA

Bradley Duncan
Lincoln Laboratory Massachusetts
  Institute of Technology
Lexington, Massachusetts, USA

Edwin Elston
University of Dayton Research Institute
Eglin Air Force Base, Florida, USA

Chris Funke
Plazmod Technologies Inc.
New Haven, Connecticut, USA

Austin Good
Samtec
Newark, Delaware, USA

Mary Herndon
Raytheon UMass Lowell Research
  Institute
Lowell, Massachusetts, USA

John Hornick
Sciperio
Phoenix, Arizona, USA

Zachary Larimore
Samtec
Newark, Delaware, USA

Nathan Lazarus
University of Delaware (formerly of US
  Army Research Laboratory)
Newark, Delaware, USA

Eric MacDonald
University of Texas at El Paso
El Paso, Texas, USA

Neal Magdefrau
Electron Microscopy Innovative
  Technologies LLC
East Hartford, Connecticut, USA

Lorenzo Migliorini
Centro Interdisciplinare Materiali
  e Interfacce Nanostrutturati
  (CIMAINA)
Interdisciplinary Center for
  Nanostructured Materials and
  Interfaces and Department of Physics
University of Milan
Milan, Italy

Paolo Milani
Centro Interdisciplinare Materiali
  e Interfacce Nanostrutturati
  (CIMAINA)
Interdisciplinary Center for
  Nanostructured Materials and
  Interfaces and Department of Physics
University of Milan
Milan, Italy

Matthew Montgomery
Chemical Engineering
Yale University
New Haven, Connecticut, USA

Andrew S. Morgan
Mechanical Engineering and Materials
  Science
Yale University
New Haven, Connecticut, USA

Shomeek Mukhopadhyay
Chemical Engineering
Yale University
New Haven, Connecticut, USA

Bhargavi Mummareddy
Youngstown State University
Youngstown, Ohio, USA

Clayton Neff
University of Dayton Research Institute
Eglin Air Force Base, Florida, USA

Spencer Nguyen
Florida State University
Tallahassee, Florida, USA

Paul Parsons
Samtec
Newark, Delaware, USA

Eleanor Rogenski
Youngstown State University
Youngstown, Ohio, USA

Eduardo A. Rojas-Nastrucci
Embry-Riddle Aeronautical University
Daytona Beach, Florida, USA

Tommaso Santaniello
Centro Interdisciplinare Materiali
  e Interfacce Nanostrutturati
  (CIMAINA)
Interdisciplinary Center for
  Nanostructured Materials and
  Interfaces and Department
  of Physics
University of Milan
Milan, Italy

Wayde Schmidt
Mat-Chem LLC
Pomfret, Connecticut, USA

Amanda Schrand
Air Force Research Laboratory
Eglin Air Force Base, Florida, USA

Evyatar Shaulsky
Chemical Engineering
Yale University
New Haven, Connecticut, USA

Paul Sheedy
Raytheon Technologies Center
East Hartford, Connecticut, USA

Chengying Xu
North Carolina State University
Raleigh, North Carolina, USA

1 Introduction to Printed Electronics

John Hornick
Sciperio

Some use the term "3D printing", while others say "additive manufacturing". Although these terms are used interchangeably in this book to discuss particular manufacturing techniques and steps in electronics manufacturing, my view is that this book is about something much more, which I call 3D manufacturing. Since its invention in the latter part of the 20th century, 3D printing has been about the layered manufacturing of parts. Although there will probably always be a demand for 3D-printed parts, and although that demand will grow and supplant some traditional manufacturing, the ultimate goal of 3D manufacturing is to produce fully functional products, not just custom, static parts. This book provides insight into recent progress in 3D manufacturing of electronic products.

"3D printing" is a collective term for several different layered-building techniques. Systems that build with these techniques generally only "print" one material at a time using a single process such as material extrusion, powder bed fusion, material jetting, and vat photopolymerization; these techniques are usually not combined in manufacturing systems. Such single-technology systems generally don't incorporate traditional manufacturing techniques (which some call "subtractive"). But when they do, they are often called "hybrid" systems because the term "3D printing" is not adequate. I call the processes performed by hybrid systems "3D manufacturing", which is not only the next step beyond 3D printing but also the probable future of manufacturing itself. Apply 3D manufacturing to electronics, and you have this book.

As I discuss in my book, *3D Printing Will Rock the World* [1], 3D printing will not replace traditional manufacturing. Nor is 3D printing a fad to be outlived by traditional manufacturing. The future of manufacturing will make the most efficient use of both 3D printing and traditional manufacturing to make products that could not be made by either 3D printing or traditional manufacturing alone. Perhaps the most powerful application of 3D manufacturing is the making of electronic devices.

Until fairly recently, electronic devices were rather formulaic: some kind of a box containing electronics. Printed circuits inside the box have been mostly planar and nonflexible. Most circuits have been efficiently fabricated only by mass production techniques. The use of the device has been largely dependent on the shape and size of the box. The resiliency of such devices to temperature, moisture, shock, pressure, and other environmental and operational conditions has been mostly dependent on the box and associated packaging materials. The 3D manufacturing of electronics

DOI: 10.1201/9781003138945-1

1

promises to make electronics, and therefore electronic devices, less dependent on the box in every way. The contributors to this book are thinking outside that box.

Making full use of traditional electronics manufacturing and taking advantage of 3D manufacturing, hybrid electronics are printed or placed and mounted on flexible or complex substrates, depending on the current state of the art, which is always advancing. The ideal applications for hybrid electronics share a common need: devices need to be resilient in harsh environments or operating conditions. They not only need to survive but also to continue to operate optimally no matter what nature or the user throws at them.

Hybrid electronics can be flexible, customized, and complex in shape (both internally and externally) and embedded in the structure or housing of a device. For example, antennas can be printed on or embedded in doubly complex surfaces, such as a device that contours to the user's hand, or flexible sensors can be integrated into devices applied to the human body. Interconnects are printed instead of etched, and integrated circuits are placed and mounted as thinned or bare die, thereby enabling smaller form factors, greater flexibility, and lighter weight, with obvious advantages for applications among the usual suspects: medical and health devices, aerospace, automotive, communications devices and other consumer electronics, UAVs, smart munitions, and sensing devices. The list goes on and is inherently incomplete. The 3D manufacturing of hybrid electronics, which consist of both printed and traditional components, will pave the way for devices that don't yet exist.

As explored in this book, the capabilities and advantages of 3D manufacturing, namely, substantially faster and cost-efficient prototyping, reduced manufacturing waste, freedom from design-for-manufacture limitations and the ability to execute highly complex and highly customized multi-material designs, the use of conductive materials, the ability to place components that cannot yet be printed, surface smoothing, printing of materials arbitrarily in three-dimensional spaces (not just XYZ), cost-effective small production or one-off production, and mass customization are supercharging the design and development of resilient hybrid electronics (RHE) for use in harsh environments or operating conditions.

There is a lot to this book. Chapter 2 starts out by asking the question "Which Printed Conductive Inks and Interconnects Survive High G and Thermal Cycling?" Dr. Schrand of the Air Force Research Laboratory Munitions Directorate and her colleagues have been instrumental in developing rigorous testing and experimental protocols to advance proof-of-concept RHE. Their insights are followed by several chapters addressing particular 3D-manufactured RHE components. Chapter 3 describes 3D-printed antennas, which are crucial to the functionality of any RHE device that communicates. Eduardo Rojas approaches this topic in his role as the Director of the WiDE Lab (Wireless Devices and Electromagnetics Laboratory) at Embry-Riddle Aeronautical University. Cheryl Xu, associate professor in the Department of Mechanical and Aerospace Engineering at NC State University, addresses printed pressure sensors for extreme environments in Chapter 4.

Other chapters explore particular 3D manufacturing technologies for the RHE sphere. In Chapter 5, Nathan Lazarus of the US Army Research Laboratory discusses metallization of 3D-printed devices. Chapter 6 is about printing electrically conductive patterns on polymeric and 3D-printed systems authored by Paolo Milani, director, Interdisciplinary Center for Nanostructured Materials and Interfaces at the University of Milano Via Celoria. A team of researchers from Yale, including small business spin-off Plazmod Technologies Inc, the Applied Physics Group at Raytheon Research Center, Raytheon UMass Lowell Research Institute, and the University of Connecticut Complex Fluids Laboratory write on the topic of direct write-printed electronics using non-equilibrium plasma-based techniques in Chapter 7. Pedro Cortes, associate professor of Chemical Engineering at Youngstown State University, and colleagues examine ceramics and embedded conductors for high-temperature applications in Chapter 8. Chapter 9 explores lifecycle challenges of printed electronics, the author of which, Kenneth Blecker, explores the topic with the perspective and experience of an engineer in the US Army Armaments Center. Lastly, this book concludes with Chapter 10 on "Three-dimensional Functional RF Devices Enabled through Additive Manufacturing" by a group of researchers at Samtec, Inc., a major global supplier of electronic components.

RHE are at the leading edge of 3D-manufactured electronics. Together with many other researchers and companies, the authors composing this book are driving the design and development of novel proof-of-concept electronic devices. The future and assured success of this field depends on the interdisciplinary interaction of many factors, including design, materials (including conductive materials and materials that adapt or become more resilient in harsh environments), 3D printing technologies and their combination into single platforms, combining 3D printing techniques with traditional manufacturing techniques, the ability to print multiple materials, and software design. As you will see, this book is an important step toward fully 3D-manufactured RHE. If you are reading this book, you may be part of the RHE revolution.

REFERENCE

[1] *3D Printing Will Rock the World.* https://smile.amazon.com/3D-Printing-Will-Rock-World/dp/1516946790/ref=sr_1_1?dchild=1&keywords=3d+printing+will+rock+the+world&qid=1614386605&sr=8-1

2 Which Printed Conductive Inks and Interconnects Survive High G and Thermal Cycling?

Clayton Neff and Edwin Elston
University of Dayton Research Institute

Amanda Schrand
Air Force Research Laboratory

2.1 INTRODUCTION

Prototyping electronic components and devices via additive manufacturing (AM) has significant advantages over traditional electronics when it comes to rapid design, customization, complex geometry (i.e. printing on conformal and irregular surfaces) and low volume production runs to generate proof of concepts, accelerate product development and enable iterative evaluation toward performance enhancement and optimization [1–11]. Printed electronics have even been called out as part of the Industrial Revolution of the Digital Age [1,10]. Typically, prototypes will contain both additively manufactured portions via 'electronics printers' as well as 'hybridization' with commercial off the shelf (COTS) components, which are integrated manually or by 'pick and place' systems. This advantageous combination of AM conductive traces, sensors, or other printed electronics paired with traditional components and interconnects is envisioned to supersede the manufacturing and performance limitations of traditional electronics through reduced size, weight, power, and cost (SWaP-C) requirements. For example, the miniaturization, materials and design customization, and relocation of sensors, including conformal antennas, with customized interconnects, broadens novel detection, signal processing, and survivability in increasingly extreme environments [12–15]. However, standard equipment, materials, and methods have not yet been vetted by the printed electronics community.

Commercial tabletop electronics printers can come in 2- to 5-axis systems with micro-dispensing or aerosol jet print head options. Commercial offerings are from companies such as nScrypt, Optomec, IDS, Inc., Nano Dimension,

DOI: 10.1201/9781003138945-2

Neotech, and Hyrel [16–21]. In the case of micro-dispense printers, the limitations remain for precision ink deposition and standoff distance where there can be risk of damaging the printing tip when printing onto objects with curvature irregularities. For example, the printing tip may come as close as 50 μm from the object's surface depending on tip diameter. In contrast, an aerosol jet print head can tolerate a much larger dispense gap standoff in the range of 1–5 mm. In addition to differences in print heads, electronics printers can be outfitted with a variety of tools for imaging or performing tests such as a scratch adhesion test (SAT) tool.

In order to print electronics onto conformal, irregular, doubly curved, or hemi-spherical object surfaces for three-dimensional (3D) manufacturing, multi-axis toolpaths are required. However, software improvements and user education are still areas of development that are being pursued by experts in Industry and Academia. At the writing of this chapter, there is no 'gold standard' software or method to use for multi-axis electronics printing or software for 3D manufacturing. Some users take the approach to write their own custom code, which provides the flexibility to shift between different printer interfaces as opposed to using vendor-specified software tools. However, generating custom code is typically performed by a user with significant programming and coding experience and can be a tedious process. Alternatively, each printer vendor provides associated software for their machine that may be limited in application. In addition, software developments continue to be upgraded and offered including nStudio software, Rhinoceros, Grasshopper, and companies like Kraetonics, LLC striving to enable greater complexity in 5-axis printing operations, conformal mapping algorithms, and 3D manufacturing.

Three aspects of printed electronics to achieve resilient, survivable, and functional hybrid-printed electronics include: (1) adhesion to the substrate for mechanical anchoring, (2) maintenance of DC or RF conductivity for sustained performance, and (3) ability to interconnect with traditional electronic circuit board components. Of course, there are many other aspects to creating novel hybrid electronic devices, but this work emphasizes that adhesion, DC & RF conductivity, and interconnect capability are co-dependent. If one of these three aspects fails, then the entire device or system is at risk for failing. Figure 2.1 delineates the dependence of these three aspects in generating viable electronics for the next generation of hybrid devices.

This chapter provides a primer on the materials and methods on ink and interconnect evaluation for resilient, survivable, and functional printed electronics. In addition, the chapter covers examples of the Development of Testing Protocols for Ink Adhesion and Impact Resilience, followed by Evaluation of Adhesion, RF Performance and Interconnects Subject to Harsh Environments, and a final section on Conclusions and Future Outlook. It is the authors' hope that elucidating the work from our lab will motivate near-term solutions, interest, and research investments into the mainstream production of printed electronics.

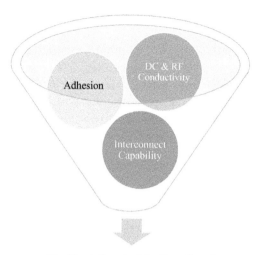

**Resilient, Survivable, Functional
Hybrid Printed Electronics**

FIGURE 2.1   Codependency of adhesion, DC and RF conductivity, and interconnect capability to create resilient, survivable, and functional hybrid-printed electronics. In addition, these three aspects must all be maintained during and after exposure to harsh environmental conditions to demonstrate device viability.

2.2   DEVELOPMENT OF TESTING PROTOCOLS FOR INK ADHESION AND IMPACT RESILIENCE

Standards, testing protocols, and qualifications give users confidence that a device or service will function efficaciously. Otherwise, users will be skeptical based on the lack of validation and verification. As a rather new field, additively manufactured electronics are lacking a set of rigorous parameters upon which to vet end product quality compared to traditionally manufactured electronics American Society for Testing and Materials (ASTM) standards. For example, there are standards for adhesion using a scratch and tape method [22–24]; however, such tests are highly operator-dependent and can be based upon manually applied force. To circumvent operator-to-operator variances, it is anticipated that the existing protocols and procedures will be assessed for not only applicability to AM, but also to be revisited and potentially revised for traditional electronics, thereby benefiting standards development overall. This section provides some examples of developments in the basic mechanical testing of adhesion for thin film materials (including conductive inks) with a custom device and harsh environmental testing of conductive inks. Furthermore, the following design and results present the effectiveness of utilizing the semi-automated SAT tool to improve the repeatability of manual cross-hatch scratch testing to nullify operator variance and experience. Another motivating factor is to provide a repeatable scratch method to understand if poor adhesion is due to a lack of adhesion itself instead of interpretation of manual scratching. It is worth noting here that there are several efforts directed at standards development for flexible, hybrid-printed electronics at Semiconductor Equipment and Materials International (SEMI), the Institute for Printed Circuits (IPC), the ASTM and at the Printed Electronics, Energetics, Materials, and Sensors (PEEMS) Center at US Army DEVCOM Armaments Center, all of which are summarized in the following reference [25].

2.2.1 SCRATCH ADHESION TEST (SAT) TOOL

Adhesion is a critical factor for conductive inks to qualify for any application with applied forces, especially to remain functional in harsh environments [14,26–31]. However, a standardized and repeatable adhesion test method is lacking in current testing practices. This work presents the development of a standard protocol for testing the adhesion of any thin coating (including conductive inks) by designing, printing, assembling, and testing a semi-automated cross-hatch SAT tool (Figure 2.2). In contrast to manual cross-hatch scratch testing for evaluating adhesion, which can yield high operator-to-operator variance, a semi-automated method allows control of the depth, speed, and planarity of the scratch to improve the repeatability of adhesion testing. The SAT tool addresses the ideal adhesion test provisions and is a quantitative measurement. However, when cross-hatch scratch testing is paired with image processing, the adhesion measurement yields a semi-quantitative value that can be used for rapidly screening materials or comparison between material sets. Additionally, the SAT tool can be readily adapted into multi-tooled manufacturing systems and is broadly applicable across the adhesion testing community [32].

2.2.1.1 Scratch Adhesion Test (SAT) Tool

The SAT tool (Figure 2.2) design incorporates an upper and lower component to allow the lower component to self-align by pivoting, if necessary, when contacting the substrate with the axle bearings. This allows for the lower SAT tool

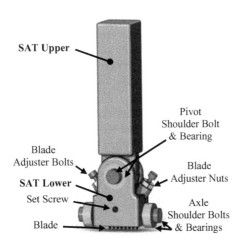

FIGURE 2.2   Diagram of scratch adhesion tester (SAT) assembly [32].

component to be normal to the scratch intended surface in case the substrate is not perfectly level or on an inclined surface. The blade should puncture the substrate with enough depth to allow the axle bearings to fix the roller depth independent of applied force. They then act as rollers and allow a smooth and consistent translation for the scratch. The blade adjuster bolts (Figure 2.2) are utilized to fine tune the planarity of the blade for uniform scratch depth of the individual blade teeth. Tightening or loosening the blade adjuster bolts will adjust the blade by contacting the corner of the blade at ~145° through the channels in the lower SAT tool.

To achieve repeatable measurements between operators or labs with different SAT tools, they will first need to be calibrated by scratching and adjusting the blade with the blade adjuster bolts to equivalent scratch depths or widths depending on which is easier to measure with available equipment. In this work, the calibration process was performed by iteratively scratching a coating, adjusting the blades for planarity, and measuring the scratch widths for each blade tooth. The scratch widths were measured with image processing tools and adjusted until uniform widths were achieved for the individual blade teeth. Scratch widths or depths can also be found with other equipment with high enough resolution to render calibration measurements, for instance, optical profilometry. The specific scratch depths and resulting widths will be material specific as the depth needs to be large enough to penetrate through the entire coating and into the substrate while also permitting the axle bearings to contact the substrate for smooth translation while scratching. If a variety of coatings are examined with uniform thickness, the scratch depth will be uniform, as well, if the blade is allowed to penetrate the substrate to the point of contacting the axle bearings/rollers.

An adjustable blade permits a variety of coatings to be tested since the depth of scratch can be increased or decreased depending on coating thickness. After fine tuning the planarity and depth of scratch for uniform scratching with preliminary testing, the SAT tool provides a quick and inexpensive adhesion testing device that reduces the deviations due to manual scratching and can be compared across labs. Even finer analysis of adhesion can be compared if the SAT tool is paired with image processing to output binary images of the cross-hatch scratch. Based upon the details below, it is envisioned that such a tool could be adapted to various electronics printers as a tool for standardizing in situ adhesion assessment.

2.2.1.2   Materials and Methods

Acrylonitrile-butadiene-styrene (ABS) substrates were machined to $40 \times 40\,mm^2$ with a 3 mm radius fillet on each corner of the square substrate from a $12 \times 12 \times 1/16$ inch^3 sheet purchased from McMaster-Carr, Part: 8586K151. Krylon Fusion (for plastic) Satin White Spray Paint was selected as the coating in this work. The Krylon spray paint was used as an inexpensive coating to preliminary test and compare against manual cross-hatch scratch testing on ABS substrates for repeatability analysis. The Krylon spray paint showed excellent adhesion, which is necessary to fine tune the blade of the SAT tool as the scratches need to be repeatable between samples. The spray paint was dispensed to a thickness of ~50 μm within an area of $20 \times 20\,mm^2$ in the center of the ABS substrates that was outlined with tape to only spray the $20 \times 20 \, \text{mm}^2$ area. The number of spray paint passes and traverse speed were held as constant as possible for a manual painting process and done in large batches to mitigate thickness variation.

2.2.1.3 Testing Procedures

1. Position the sample flat and fixed to a stage with at least 1D motion. To adhere the sample to the stage, double-sided tape works well or use a printed fixture that sample press-fits into for repeatable positioning and quick and easy sample removal. The printed fixture has a pocket with 1/16" tall walls that has dimensions a couple hundred microns larger than the ABS substrates to constrain the ABS substrate from moving while running the rest of the test procedures.
2. Position SAT tool (mount if not already) where the blade is a few millimeters ahead of the film or coating to be scratched.
3. Run SAT tool script (pseudo code in the appendix).
   a. Lower to scratch position with enough contact pressure for desired scratch depth (needs to be determined in the initial installation and fine tuning of the scratch). In this work, 95 psi was supplied to the pneumatic cylinder in the electronics printer when actuating the SAT tool.
   b. Scratch with 25 mm of translation or whatever distance that will scratch through entire film or coating with a suitable velocity. A value of 2.5 mm/s was chosen for preliminary testing.
   c. Lift SAT tool.
   d. Reset to starting position.
4. Rotate the sample 90° clockwise.
5. Run the SAT tool script.
6. Remove the sample.
   a. Cross-hatch SAT tool testing is completed.
   b. Inspect with microscope and/or pair with image processing for analysis.
7. Repeat steps 1–6 for more samples.

The SAT tool can be mounted in an electronics printer with 95 psi supplied to the pneumatic cylinder, which induces 7.4 lbs to the blade or 0.93 lbs/tooth. After fine tuning in the iterative calibration procedures of scratching, imaging/measuring, adjusting, each scratch width from the multi-tooth blade was $150 \pm 8 \, \mu\text{m}$ in an initial repeatability study. Image processing provided binary image analysis of the cross-hatched adhesion patterns after scratching and calculated the scratch widths and percentage of ink remaining post-test as white area or '% W'. The appendix includes the assembly, installation, hardware, pseudo code, and CAD drawings for the SAT tool.

To compare the effectiveness of the SAT tool vs. manual cross-hatch scratch testing, five different operators were each given an ABS/Krylon spray paint sample, the manual cross-hatch scratch testing tool from Gardco Inc., and instructions from ASTM F1842 for Determining Ink or Coating Adhesion. Prior to manual cross-hatch scratching the ABS/Krylon paint sample, the operators were asked to practice cross-hatch scratching a spare ABS substrate three separate times to get a feel for manually scratching. Further practice was not requested as operator variance is a variable in this study.

2.2.1.4   Results

Figure 2.3 depicts the manual scratches between the five different operators, which show a wide range of the Krylon spray paint remaining, 85.84% W to 92.65% W, with an average and standard deviation of 92.05 ± 1.95%. The manual scratches have a relatively high degree of variance due to inconsistent pressure from not maintaining constant pressure throughout the scratch, a non-level blade resulting in deeper scratches on one edge than the other and misaligned scratch patterns (Figures 2.3 and 2.6). These inconsistencies emphasize the challenges of manual scratching as operator experience becomes valuable, but the variance will still exist even if mitigated.

Conversely, the SAT tool scratches in Figure 2.4 show much more uniform scratching with an average of 87.70 ± 0.56% white remaining. This results in ~4× smaller deviations for SAT tool testing and allows finer differences in scratch testing to be detected. Adhesion measurements with a SAT tool also nullify the value of operator experience as once initial calibration is completed, all operators will achieve equivalent results with the semi-automated device. The utilization of a semi-automated SAT tool also provides a tool that bypasses the interpretation aspect of manual scratching

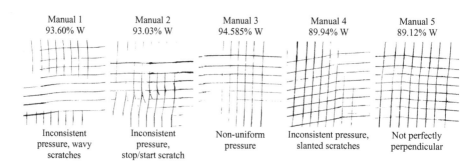

FIGURE 2.3   Manual cross-hatch scratches with five different operators. Note the non-uniform scratches from inconsistent pressure and other manual defects. Average white remaining after scratching 92.05% ± 1.95% [32].

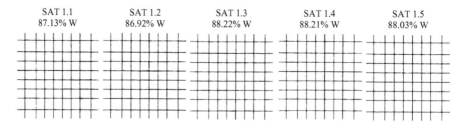

FIGURE 2.4   SAT cross-hatch scratches. Note the repeatability in the scratches. Average white remaining after scratching 87.70% ± 0.56% [32].

and the results are more purely indicative of the adhesion without questioning if poor adhesion is due to a lack of adhesion or lack of quality manual scratching. Therefore, the SAT tool provides a valuable tool for finer comparison of cross-hatch scratch testing by controlling the depth, speed, pressure, and planarity with a semi-quantitative adhesion measurement.

Qualitative adhesion measurements such as manual cross-hatch scratch testing and scotch tape testing can be advantageous for quick and inexpensive adhesion testing, but also include uncontrolled variables that make repeatable testing a challenge. Uncontrolled variables include speed, pressure, planarity, and depth, which may all vary between tests and even more likely between operators. This makes comparing adhesion a coarse indication, and finer analysis between materials and labs may be prohibited. This work shows that the variance between cross-hatch scratch testing can be reduced by a factor of ~4x with the design, printing, and testing of a semi-automated cross-hatch scratch adhesion test (SAT) tool. The SAT tool provides uniform scratches that overcome manual scratching inconsistencies such as non-uniform scratches and stop/start scratches by controlling the speed, planarity, and depth of the scratch. The semi-automated SAT tool nullifies operator variance and provides finer analysis for comparisons between materials and labs. Repeatable SAT tool testing also bypasses the interpretation aspect of qualitative adhesion measurements as the material removed due to scratching is from a lack of adhesion and not from inexperience with manual scratching. The combination of image processing with cross-hatch scratch testing also provides an effective method for semi-quantitative adhesion measurements. Therefore, a semi-quantitative and repeatable adhesion test method for conductive inks with a SAT tool offers a standard protocol to be adopted to inspire trust in conductive ink adhesion for printed electronics.

2.2.2 High G Testing

Conductive inks and other thin films are difficult to test mechanically under applied forces. Adhesion tests offer the ability to capture a form of strength, however, they are limited to mostly a static test of interfacial shear strength. High g testing offers a dynamic mechanical test method to evaluate the resiliency of conductive inks and thin films to endure significant forces in a harsh environmental test. Furthermore, high g testing and adhesion testing can be paired to evaluate the impact of high g testing on adhesion strength. For example, a set of conductive ink samples can be adhesion tested as prepared, while another set of samples with the same processing are adhesion tested after being subjected to high g testing. If the adhesion performance of the set of samples exposed to high g is worse, then it reveals that adhesion is compromised at that level of dynamic testing and applied forces. The following paragraph presents data utilizing a MIL STD 883F Method 2002.4 Mechanical Shock high g test for printed electronics [33].

High g (acceleration) testing utilized an MTS drop tower (unless otherwise noted) to induce mechanical accelerations up to 50,000 g's with a pulse width of 0.1 ms, which follows the profile of MIL STD 883F Method 2002.4 Mechanical Shock Condition F and more severe than condition G. The typical acceleration curves can

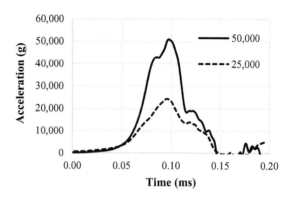

FIGURE 2.5   High g testing according to MIL STD 883 [15].

be found in Figure 2.5 with a 25,000 and 50,000 g waveform. After each test, the maximum acceleration and pulse width are recorded, followed by an evaluation for the harsh environmental high g test to access survivability to dynamic forces. The post high g test to access survivability includes checking electrical continuity and/or antenna performance for electrical test samples and cross-hatch scratch and tape test for mechanical (adhesion) test samples.

2.2.3   SUMMARY OF TESTING PROTOCOLS FOR PRINTED ELECTRONICS

A lack of standards and protocols for mechanical testing of conductive inks remains an obstacle for rapid adoption of printed electronics into mainstream manufacturing. This work offers both a static and dynamic test to evaluate the efficacy of printed electronics to withstand a range of applied forces. The evaluation of adhesion can be accomplished with a SAT tool for repeatable and semi-quantitative comparison of adhesion before and after other forms of testing as well as to mitigate user-to-user variances with traditional tape testing procedures. However, adhesion testing alone only offers a static mechanical test of interfacial shear strength. By contrast, high g testing offers a method to examine the response of conductive inks and other thin films for their 'survivability', which is a particularly useful assessment for components exposed to the harsh environments encountered in the defense industry. Adhesion and high g testing can also be paired to assess the mechanical survivability of conductive inks and thin films with visually straightforward images that show material remaining indicative of good adhesion.

Although the examples provided here are in the form of a simple printed conductive ink 'patch' or antenna (design not shown), the design should remain simple for the purpose of generating standardized protocols and procedures so they can be adopted in various machines and labs. Similarly, the adhesion tool could be customized for different printers and the mount for the high g testing could be customized for the machine being utilized. For some purposes, simple lines or serpentine patterns are useful for standards development dependent upon the properties of the printed electronics that are being examined. Further, generating standards that can be used on different printers to assess similar properties (i.e. line widths, thickness, conductivity), as well as lab-to-lab and materials' comparisons, all contribute to increasing the confidence and adoption of printed electronics into mainstream use.

2.3   EVALUATION OF ADHESION, RF PERFORMANCE, AND INTERCONNECTS SUBJECT TO HARSH ENVIRONMENTS

In the following sections, the details will include both micro-dispensed and aerosol jet conductive inks utilizing the protocols for adhesion and high g testing that were explained in the previous section to evaluate the three aspects of adhesion, DC & RF conductivity, and interconnect capability for resilient, survivable, and functional hybrid-printed electronics with regard to surviving harsh environments. Together these results provide suggested materials selection protocols to support future development efforts.

2.3.1   Micro-Dispensed Conductive Inks

In this section, micro-dispensed inks were evaluated for studies of adhesion, RF performance, and electrical interconnects and include DuPont CB028, KA801, CB230, and Novacentrix 57B. Previous work shows the microstructure and process characterization of micro-dispensed ink [34], and utility/variety of printed electronics that can be produced employing the micro-dispensing technique including RFID and coplanar waveguides [35–39], vertical interconnects [40], non-conformal and conformal antennas [41–45], and laser-enhanced micro-dispensed electronics for improved performance [46–49].

2.3.1.1   Adhesion Evaluation of Micro-Dispensed Ink before and after Harsh Environments

In the characterization methods covered in this chapter, adhesion is the first property measured due to inherent disruption of printed electronics function without adequate adhesion. An evaluation of adhesion for two commercially available micro-dispensed conductive inks (DuPont CB028 and KA801) printed onto a radar transparent substrate (polyether ether ketone, PEEK) showed that CB028 fares better than KA801 during adhesion testing after exposure to high accelerations up to 20,000 g and temperature cycling from −54°C to +71°C [14]. Note that the high g testing performed in this section was on planar substrates with the impact forces being perpendicular to the plane of the 20 × 20 mm printed ink patches under evaluation.

Figure 2.6 shows binary images of printed ink patch samples before and after exposure to high g and/or thermal cycling, then assessed for remaining material or adhesion via a cross-hatch scratch and tape test. The top row of samples (Figure 2.6a–d) are representative of DuPont CB028 conductive ink and contain up to 87% ink coverage even after being subjected to thermal cycling and high acceleration mechanical shock. Conversely, the bottom row of samples (Figure 2.6e–h) are representative of DuPont KA801 conductive ink and show a greater amount of ink removal after the

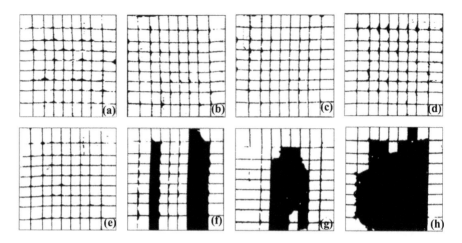

FIGURE 2.6 Binary images of samples after adhesion testing: (a) CB028 as-printed, (b) CB028 thermal, (c) CB028 high g, (d) thermal then high g, (e) KA801 as-printed, (f) KA801 thermal, (g) KA801 high g, and (h) KA801 thermal then high g [14].

combined thermal and high g testing conditions (Figure 2.6h). Further, the use of a manual scratch and tape test for these studies results in noticeable irregularities in the scratch patterns. Therefore, a standard procedure was recommended to use a more qualitative and uniform process such as using a SAT tool for comparison of ink adhesion and impact of harsh environmental testing.

2.3.1.2   RF Performance of Micro-Dispensed Inks

In this section, the RF performance of 20 mm × 20 mm square patch antennas printed in two different conductive ink formulations (CB02 and KA801) onto PEEK polymer substrates was investigated. One of the motivating factors for this study was to assess if RF performance could be predicted by simple visual examinations after subjecting the materials to high g and/or thermal cycling as a quick and inexpensive screening tool.

Two figures of merit were extracted from the printed antenna chamber measurements: the system reflection and transmission coefficients. It is desired to maintain a low reflection coefficient and a high transmission coefficient at the antennas' resonant frequency after environmental testing. Degradation in these metrics can be evidence of changes to antenna impedance due to damage, such as delamination of the conducting ink from the patch surface.

Reflection coefficients plotted for CB028 signify a good impedance match, with all values below −15 dB at resonance (Figure 2.7a). In contrast, the reflection coefficient curves for KA801 after high g and combined thermal and high g testing show a flatline at around 0 dB (Figure 2.7b), signifying a total loss of impedance match and therefore antenna failure. For the two failure cases plotted in Figure 2.7b, physical damage to the antenna was obvious by visual inspection, as large sections of the conducting ink delaminated from the PEEK substrate. Measured transmission

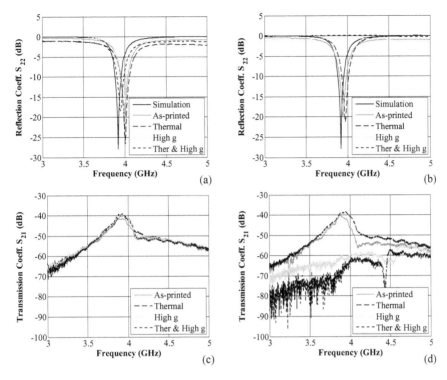

FIGURE 2.7 (a) Reflection coefficient: CB028, (b) reflection coefficient: KA801, (c) transmission coefficient: CB028, and (d) transmission coefficient: KA801 [14].

coefficients for the same eight example cases are plotted in Figure 2.7c and d. Antenna performance is once again consistent across test cases for the CB028 ink (Figure 2.7c), with a peak transmission coefficient of around −40 dB. For the KA801 ink, antenna failure is once again apparent for the two high g cases, as the transmission coefficient has dropped by around 20 dB due to delaminated blotches of ink missing, shifting antenna performance completely Figure 2.7d.

This data shows that antenna performance is maintained for CB028 test cases. In contrast, the KA801 results demonstrate far greater variability in performance. All of the high g and one of the thermal cycle test cases demonstrate a degraded reflection coefficient, and three of the high g test cases demonstrate a significantly degraded transmission coefficient. Physical damage was visually apparent in these cases, and it included the delamination of large sections of the patch antennas as well as rippling in the patch surface, which greatly hinders and shifts antenna performance. These results indicate the CB028 ink as prepared/printed into an RF antenna would allow more stable performance (i.e. less sensitive to environmental shocks) compared to the KA801 ink. Furthermore, the RF performance results of this section follow the same trends as the adhesion results of the previous section, which supports the use of adhesion testing as a rapid preliminary screening method for printed electronics subjected to harsh environments.

2.3.1.3 Electrical Interconnect Testing with Micro-Dispensed Conductive Inks

In this section, a variety of solders and conductive epoxy were assessed as interconnecting materials to printed micro-dispensing inks. The example of a square printed patch was replicated as the standard design to evaluate the effectiveness of different materials for connecting or bonding the printed conductive ink component (i.e. solder pad, antenna, sensor) to traditional circuit components (pin, circuit board, etc.). DuPont CB230 copper and Novacentrix 57B silver conductive were chosen as the printed inks due to their ability to be soldered. The interconnecting materials included both traditional leaded solder and low-temperature solders as well as EPO-TEK H20E conductive epoxy. Table 2.1 presents the variety of interconnect materials studied. Previous work demonstrates the effectiveness of interconnecting materials as conductive epoxies [50,51], low-temperature silver pastes [52–55], inkjet-printed interconnections [56,57], low-temperature solders with various compositions of tin, bismuth, and indium [58–71]; however, there is a gap for harsh environmental testing of conductive epoxy and low-temperature solders on printed micro-dispensed ink. Although some work can be found on interconnections for high-temperature electronics packaging and thermal cycling [72–76], this work focuses on wire bond pull testing and high g testing to evaluate the effectiveness of conductive epoxy and low-temperature solders for resilient/survivable hybrid-printed electronics.

The effectiveness of the bond of interconnect materials to printed ink squares was examined with wire pull bond testing and high g's up to 50,000 g's of mechanical acceleration [15]. Wire bond pull samples were fabricated by interconnecting the solders or H20E conductive epoxy to a 0.81 mm outer diameter (20 gauge) 100 mm long braided copper wire to the printed conductive ink, as shown in Figure 2.8. Figure 2.8 also shows a high g sample with a 4×6 array of high g interconnect samples for a total of 24 interconnect samples per $45 \times 30\,mm^2$ PEEK substrate. In this case, the top two rows consist of H20E conductive epoxy interconnect samples, while the bottom two rows are low-temperature Indium, Inc. solder 97In/3Ag.

Wire bond pull testing results show that conductive epoxy has a higher maximum interfacial stress (Figures 2.9a and 2.9b); however, it is more brittle than the low-temperature solders when using a solvent-free flux remover (Figures 2.9c and 2.9d).

TABLE 2.1
Traditional and Low-Temperature Solders and Conductive Epoxy for Electrical Interconnect Testing [15]

Interconnect Material	Composition (wt%)	Solder Tip Temperature (°C)
Sn/Pb 'Eutectic' solder	63Sn/37Pb	370
Indium Inc. solder 281	58Bi/42Sn	245
Indium Inc. solder 282	57Bi/42Sn/1Ag	245
Indium Inc. solder 290	97In/3Ag	245
Indium Inc. solder 1E	52In/48Sn	220
EPO-TEK H20E conductive epoxy	60–100Ag/30-60epoxy	Curing: 2 hours @ 100°C

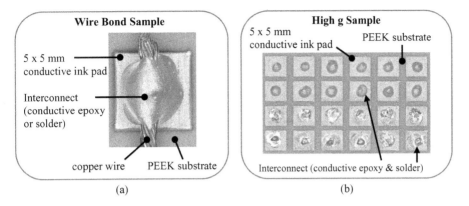

FIGURE 2.8   Diagrams of wire bond pull (left) and high g (right) array of electrical interconnect samples [15].

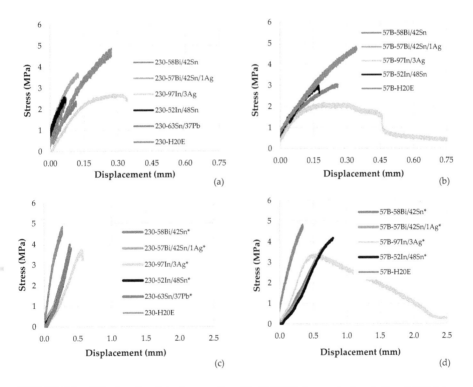

FIGURE 2.9   Wire bond pull testing results with CB230 inks graphed on left and 57B inks on right. The bottom graphs denoted with '*' designate a solvent-free flux remover method for the low-temperature solders compared to top graphs where a solvent-based flux remover was used. A solvent-free flux remover increases the ductility of the wire bond pull testing samples [15].

Low-temperature solder 97In/3Ag shows high ductility (Figure 2.9)—as expected with a high weight composition of soft and ductile indium. Low-temperature solder 97In/3Ag was also the only interconnect material to have a mixed failure mode with wire pullout. It is noteworthy when soldering to a conductive ink, the use of a solvent-free flux remover method (Figure 2.9, lower graphs) should be utilized as when comparing to the upper graphs where a flux remover with solvent(s) is shown to have a deleterious effect on the adhesion of the conductive ink.

High g testing data from Table 2.2 shows both H20E conductive epoxy and 97In/3Ag low-temperature solder perform well as interconnect materials subjected to survive harsh environments with ~40 mΩ of resistance change when exposed up to 50,000 g's. Conductive epoxy may be more preferential for printed electronic applications as: (1) it is more feasible to integrate into a multi-tooled AM process, (2) it does not have the potential for formation of intermetallic compounds that could embrittle/weaken the interconnect during thermal cycling from normal electronics operation, and (3) it is less dense and therefore has reduced mass and resulting forces when subjected to high accelerations, which overcomes the brittle nature of the conductive epoxy. The silver conductive ink examined here (CB028) shows better performance than copper conductive ink (CB230) when soldering and this is likely due to the formation of intermetallic compounds, which can contribute to an embrittled interconnection.

The wire pull bond testing presented here is one method to assess interconnect bonding to printed inks for early stage materials selection onto different substrates. Wire bond pull testing can also be implemented to examine the effect of high g testing or other harsh environmental conditions by performing wire bond pull testing after being subjected to harsh environmental conditions. The applicability of this protocol could be related to applying a small amount of conductive epoxy or solder to attach an RF cable center pin to the surface of a printed patch antenna. However, printed connections, when soldered on to conductive ink with traditional Sn/Pb solder, tend to vaporize the heat sensitive inks, leading to a degradation in the connection. Therefore, traditional methods need careful evaluation before they are applied to AM electronics, and materials should be evaluated on a case-by-case basis.

2.3.2 Aerosol Jet Conductive Inks

An IDS, Inc. aerosol jet system was also utilized to evaluate jettable conductive inks including UT Dots Ag40x and ChemCubed Ag1037 for survivability to harsh environments with evaluating the aspects of adhesion and electrical interconnects similar to the test methods employed in the previous section. PEEK sheets (McMaster-Carr Part: 8504K71) were typically used for survivability studies unless otherwise specified. Differing from micro-dispensing, aerosol jet printing produces a finer edge quality (greater precision) and a thinner deposition (less mass) of approximately half a micron (0.5 μm) per layer depending on print parameters. Therefore, multiple layers are typically applied when utilizing the aerosol jet conductive deposition process as the skin depth for printed electronics to maintain functionality (i.e. high frequency conductivity) becomes a concern. Similar to micro-dispensing, the IDS Nanojet™ still requires careful materials selection, evaluation of ink adhesion to various substrates,

TABLE 2.2
High g Testing Results for CB230 (Cu-based) and CB028 (Ag-based) Conductive Inks Connected to a Copper Wire via Various Solders or Conductive Epoxy Materials [15]

Interconnection	Area (mm²)	Volume (mm³)	V/A (mm)	Max H (mm)	Ro (mΩ)	25,000 High g		50,000 High g	
						ΔR (mΩ)	OC	ΔR (mΩ)	OC
57B-97In/3Ag	10.36 ± 1.64	2.09 ± 0.59	0.20	0.75 ± 0.17	3.1 ± 1.0	0.4 ± 0.8	0/24	37.5 ± 78.0	2/24
CB230-97In/3Ag	9.73 ± 0.66	3.30 ± 0.52	0.34	N/A	1.7 ± 0.4	11.0 ± 2.4	0/6	3,000 ± 2,000	2/6
CB230-63Sn/37Pb	7.16 ± 0.25	1.44 ± 0.12	0.20	N/A	2.3 ± 0.2	41.0 ± 25.2	1/6	5,000 ± 3,000	1/6
CB230-H20E	5.02 ± 0.64	1.30 ± 0.13	0.26	N/A	8.3 ± 1.5	122.6 ± 201.8	0/6	9,000 ± 4,000	2/6
CB028-H20E	5.57 ± 0.30	1.66 ± 0.13	0.30	N/A	20.6 ± 5.6	201.8 159.6	1/6	784 ± 1,417	1/6

TABLE 2.3

Comparison of Micro-dispensed vs. Aerosol Jet Conductive Inks

Micro-dispensing	Aerosol Jet
• Dispense gap of 10–100 µm and larger depending on dispense tip inner diameter (ID) • Dispense gap is critical to control and remain constant, especially as the dispense gap gets smaller • Line widths greater than 100 µm are common with increasing effort for line widths < 50 µm • Line thicknesses of tens of microns • Polymer-based inks commercially available in a variety of options	• Dispense gap more flexible from 2 to 5 mm • Dispense gap is not as sensitive to maintain critical distance • Line widths below 20 µm readily achievable • Line thicknesses of a micron and below. Many layers may be required for high current, which may be a slow process • Solution-based 'particle-free' reactive Ag inks with conductivities approaching bulk Ag that are much greater than micro-dispensed inks

variations in consistency of the prints, and the requirement to generate the appropriate ink thickness for measurements. A summary comparison of the two different printing options is shown in Table 2.3, and the following references show the utility and variety of printed electronics that can be produced employing the aerosol jet technique [77–85].

2.3.2.1 Adhesion Evaluation of Aerosol Jet Conductive Inks

In this section, two aerosol jet printable inks (ChemCubed Ag 1037 with and without primer and UT Dots Ag40x) were printed into patches on PEEK substrates. The adhesion tests were set up to examine the effects of substrate surface treatment with ChemCubed Dielectric primers 7 and 8 as adhesion promoters prior to ChemCubed Ag1037 ink deposition, the application of Loctite 7471 primer as an adhesion promoter prior to ChemCubed Ag1037 ink printing with print bed temperatures ranging from 45°C to 90°C with intervals of 15°C, and UT Dots Ag40x conductive ink as deposited after being subjected to impact forces up to 25,000 g's while varying the curing temperature from 200°C to 300°C with intervals of 25°C. Adhesion tests were performed either using the SAT tool of Section 2.2.1 or manual cross-hatch scratch testing, followed by the tape removal test on PEEK substrates. All PEEK substrates were wiped with acetone, followed by IPA, immediately before adhesion promoter or ink deposition as a standard.

Figure 2.10 shows different types of adhesion failure modes depending on how much of the adhesion sample (conductive ink in this case) remains after the scratch and tape test. ChemCubed Ag1037 ink was examined for adhesive failure after different surface preparations. Surface preparation with the ChemCubed Dielectric primers 7 and 8 prior to conductive ink deposition resulted in complete dielectric and conductive ink delamination, leaving behind only the scratch marks resembling the adhesive failure mode of Figure 2.10b. Therefore, the dielectric primers are not effective as adhesion promoters for ChemCubed Ag1037 on PEEK substrates.

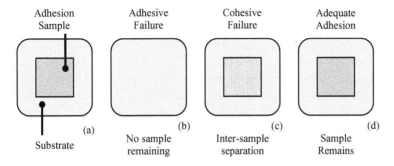

FIGURE 2.10 Adhesion failure modes: (a) adhesion sample on substrate, (b) adhesive failure of sample with no sample remaining after scratch and tape test, (c) cohesive failure of sample with inter-sample separation (in this case the ink) after scratch and tape test, and (d) adequate adhesion with none or very little of the sample being removed after scratch and tape test.

On the other hand, a single spray of Loctite 7471 primer prior to printing the ChemCubed 1037 ink shows improvement of ink adhesion compared to the dielectric primer surface treatments. However, the primer was not able to maintain adequate adhesion, and all samples exhibit inter-sample cohesive delamination after scratch and tape test. In an effort to increase ink adhesion, the print bed temperature was varied from 45°C to 90°C, which affects the curing of the ink. However, the samples still retained a moderate level of delamination resembling the cohesive failure mode of Figure 2.10c. These conditions were the author's preliminary attempts of screening ChemCubed Ag1037 for harsh environments. More aggressive surface treatments and other substrate materials may act very differently, but was not explored further in this work.

Conversely, UT Dots Ag40× conductive ink was printed for adhesion test samples and box oven heat cured to temperatures ranging from 200°C to 300°C with intervals of 25°C. After curing, the substrates were subjected to 25,000 g's prior being subjected to adhesion testing. The adhesion samples for UT Dots Ag40× were subjected to rigorous harsh environmental testing prior to adhesion testing as the authors found the conductive ink to have excellent adhesion when printed on PEEK substrates. Even after being subjected to high g testing of 25,000 g, adequate ink adhesion was achieved for UT Dots Ag40× once the curing temperature was raised above the boiling point of the ink co-solvent terpineol (219°C). Adhesion samples with a curing temperature of 225°C–300°C show excellent adhesion with the only ink removed where the cross-hatch scratch pattern was scribed and no inter-layer delamination found after tape testing, resembling closely to the adequate adhesion mode of Figure 2.10d. Therefore, the aerosol ink that demonstrated the best adhesive performance under varying surface treatment and curing conditions in these studies was UT Dots Ag40× on PEEK substrates.

2.3.2.2 Electrical Interconnect Testing with Aerosol Jet Conductive Inks

In contrast to the previous work with micro-dispensed ink interconnects, the aerosol jet conductive ink UT Dots Ag40× was only examined with conductive epoxy interconnect materials. This is due to the vaporization of aerosol conductive inks when subjected to the local heating conditions of low-temperature and conventional solders. Overall, the interconnect materials were further limited to epoxy-based polymer conductive adhesives due to their high adhesive strength. Three conductive epoxies were evaluated for resiliency to high g's for aerosol jetted conductive ink interconnections with the same configuration and testing conditions as described in Section 2.2.2 and Figure 2.8. The conductive epoxies for interconnect testing to aerosol jet conductive inks include: MasterBond EP21TDCS, MasterBond MS151S, and EPO-TEK H20E. Both EP21 and H20E consist of epoxy resin-based silver-filled conductive adhesive, while MS15 was selected as a silicone-based conductive adhesive that may provide inherently more ductility. It was hypothesized that an inherently more ductile interconnect material would be more flexible under high g loading, thus more forgiving to deformation under loading. Each conductive adhesive was prepared according to the manufacturer specifications and deposited carefully by hand to establish the interconnect from the pins to solder pads as shown in Figure 2.8.

Once the conductive epoxies were cured, the samples were exposed to harsh environmental conditions of high g according to MIL STD 883F Method 2002.4 with accelerations up to 50,000 g and a pulse width of 0.1 ms. Furthermore, the interconnect samples experienced both normal and shear forces as a 17° testing fixture was utilized to induce the mixture of forces. Resiliency and survivability of the interconnection of the conductive epoxies was evaluated with a 2-probe resistance measurements comparing the resistance before and after being subjected to high g in a similar method as previous work [15].

Key takeaways from harsh environmental testing of aerosol jet conductive ink and conductive epoxy interconnections reveal that MasterBond EP21 epoxy resin-based conductive epoxy has the most resiliency when subjected up to 50,000 g. MasterBond EP21 did not have any interconnect open circuits for the 16 samples tested. MasterBond MS15 silicon-based conductive epoxy interconnect testing up to 50,000 g only resulted in open circuits for one of eight samples tested, but had a higher overall resistance and variance after high g testing when comparing MasterBond EP21. EPO-TEK H20E on the other hand showed the worse high g resiliency for these material sets with 7 of 16 open circuits and the highest resistance after being subjected up to 50,000 g.

As previously mentioned, it was hypothesized that a silicon-based conductive epoxy would show the greatest resiliency to harsh environmental testing. However, the added ductility of the silicone could have allowed too much flexing from high g resulting in additional strain on the interconnection and increasing the resistance. Additional strain on the interconnection could cause the electrical pin to migrate from center, which will induce more strain and thus deformation on the printed ink. Both effects may lead to increased resistance of the electrical interconnection. This study reminds us of the importance to seek recommendations from manufacturers, as when examining the datasheet for MasterBond EP21, it is recommended for survivability to mechanical shock, which supports our findings in this study. Meanwhile, EPO-TEK H20E stipulates a likeliness for thermal management. In other words, the interconnect behavior may act very differently under varying harsh environmental conditions, such as thermal cycling.

Electrical interconnect testing demonstrates a feasible testing method of all three aspects of resilient, survivable, and functional hybrid-printed electronics. Adhesion of the conductive ink to a substrate, as well as the adhesion of the interconnect material to an electrical pin (COTS component), is tested. Resistance measurements of an interconnect sample provide the resiliency to maintain intended electrical performance. This supports adhesion, DC and RF conductivity, and interconnect materials need to coexist and harmonize for resilient, survivable, and functional hybrid-printed electronics.

2.3.3 Summary of Printed Electronics Materials Evaluation

Printed electronic materials have yet to be qualified for many applications, including the harsh environments encountered in the defense industry. Without requirements-based qualification testing, printed electronics will remain limited in application. Therefore, this section provided a glimpse into the further development and use of protocols and materials testing procedures for assessing printed ink: (1) adhesion, (2) resistance to high g impact, (3) thermal cycling, (4) combinations of high g impact and thermal cycling, (5) associated RF performance, and (6) interconnect strength between printed conductive inks and traditional circuit board components. Although limited in scope, the results in this section demonstrate the resiliency and survivability of certain material combinations. Together, the development of these initial protocols and materials testing procedures provides a context for printed electronics requirements starting point to outline some minimum testing requirements as the printed electronics community progresses to increase sample quality and address key limitations preventing printed electronics from being adopted mainstream.

When applying these developed protocols, it was found that both micro-dispensed and aerosol jet conductive inks can survive harsh conditions. However, after examining several different methods to prepare the samples (materials, surface treatments, varied curing temperature), there was clearly better adhesion and survivability of certain materials printed onto PEEK substrates as well as the interconnect materials. Five take-home points are provided to summarize this section: (1) DuPont CB028 micro-dispensed conductive ink exhibits adequate adhesion and resiliency to harsh environments including high g's and thermal cycling when printed onto PEEK substrates with conductive epoxy interconnects, (2) DuPont CB230 solderable micro-dispensed conductive ink exhibits adequate adhesion and resiliency to harsh environments including high g's when printed onto PEEK substrates with conductive epoxy and low-temperature solder interconnects, (3) Conductive epoxy interconnects are more resilient compared to low-temperature solders due to their lower density and thus lower resulting forces upon exposure to high g's, (4) UT Dots Ag40x aerosol jet conductive ink shows adequate adhesion and resiliency on PEEK substrates when exposed up to 50,000 g's, and (5) MasterBond EP21 conductive epoxy shows the greatest resiliency as an electrical interconnect to aerosol jet UT Dots Ag40x conductive ink.

2.4 CONCLUSIONS AND FUTURE OUTLOOK

The field of printed electronics is on the cusp of providing proof of concepts for many innovative and conformal 3D electronic devices by merging functionality and form. Despite their significant potential, the development of testing protocols to evaluate and standardize conductive ink materials and interconnects will be required based upon the intended use and environment. However, incorporating the lessons learned from both polymer and metal AM will undoubtedly influence the innovative use and mainstream adoption of printed electronics. Part of this effort will be to determine materials pairings for the best compatibility of multiple materials. In contrast to the use of conductive ink materials, the direct metal deposition onto or into polymers via (1) plasma printing or (2) sputtering or electroplating followed by laser etching may prove useful in electronic prototypes that require high metal coverage. Another aspect is the adoption of quality control and qualification efforts for printed electronics as the materials, machines, and processes for evaluation continue to mature.

Translating multi-materials printing 'out of the box' from tabletop and gantry-style printers to robotic arms is envisioned to be developed concurrently with parallel technologies including artificial intelligence (AI), machine learning (ML), and increased automation into realizing the next generation of 'smart factories'. Here, multiple print heads and machines could work autonomously to contribute to a final product in an assembly line manner. The hallmark of this concept is that the printed electronics and components would be integrated more organically within structures, and the setup would be modular in order to add or remove materials, pick and place, in situ quality controls, processing, finishing, and packaging capabilities. The use of 5- and even 6-axis machines to truly reap the benefits of electronics prototyping, and eventually manufacturing production, will require increasingly complex software and hardware as more and more tools are combined. All of this is being pursued with national security and ITAR considerations for some of the printer upgrades, such as 5-axis motion control, which is export controlled. Therefore, the interactions between the printed electronics community, consisting of printer vendors, materials suppliers, software and hardware developers, and researchers, will require close communication to shape the technology.

A substantial investment has been made in National Manufacturing Institutes, including NextFlex, which focuses on the development of flexible, hybrid electronics and was established in 2015 [86]. NextFlex's mission is to advance US manufacturing through public–private partnerships that facilitate printed electronics innovation, commercialization, workforce development, and promote a sustainable ecosystem for advanced manufacturing. Indeed, the forecast from IDTechEx for applications in the flexible hybrid-printed electronics industry continues to grow from its current $4.7 billion status to a projected $5.7 billion for 2027 and reaching $9.8 billion by the year 2032 [87]. This prediction includes technologies/ applications ranging from sensors for wearables, automotive interiors, flexible displays (i.e. electrochromic, electrophoretic), photovoltaics, batteries, and circuitry, to name a few. Therefore, it is with no doubt that there will continue to be new offerings to the markets of materials, methods, and characterization that will offer near-term solutions and motivate interest and research investments into the mainstream production of printed electronics.

ACKNOWLEDGMENTS

Funding for this research was provided by Joint Enhanced Munitions Technology Program (formerly Joint Fuze Technology Program), National Research Council Post-Doctoral Fellowship, and AFRL/SBO Small Business Office Innovation Incentive Tech Sprint awards. The authors also like to acknowledge the assistance of the Rapid Prototyping Facility on Eglin Air Force Base, laboratory technical support by Mr. Chris Kimbrough and other technical support, and discussion by Branch Technical Advisor Mr. George Jolly and technical area lead Mr. Evan Young. Cleared for DISTRIBUTION A. Approved for public release, distribution unlimited (AFRL-2023-3199).

REFERENCES

1. Berman, Barry, 3-D printing: the new industrial revolution. *Business Horizons*, 2012. **55**(2): p. 155–162.
2. Cui, Zheng, *Printed Electronics: Materials, Technologies, & Applications*. 2016, Singapore: John Wiley & Sons.
3. Espalin, David, Muse, Danny W., MacDonald, Eric, and Wicker, Ryan B., 3D Printing multifunctionality: structures with electronics. *International Journal of Advanced Manufacturing Technology*, 2014. **72**(5–8): p. 963–979.
4. MacDonald, Eric and Wicker, Ryan, Multiprocess 3D printing for increasing component functionality. *Science*, 2016. **353**(6307): p. aaf2093.
5. Schrand, Amanda, Additive manufacturing: from form to function. *Strategic Studies Quarterly*, 2016. **10**(3): p. 74–90.
6. Schrand, Amanda, Additive manufacturing in the DoD. *DSIAC Journal, Advanced Materials*, 2018. **5**(4): p. 31–38.
7. Bandyopadhyay, Amit and Bose, Susmita, Additive manufacturing. *Ch.1 Global Engineering and Additive Manufacturing*. 2016, Boca Raton, FL: CRC Press.
8. Gebhardt, Andreas, *Understanding Additive Manufacturing*. 2012, Carl Hanser Verlag: Hanser publication Inc.
9. Gibson, Ian, Rosen, David, and Stucker, Brent, *Additive Manufacturing Technologies: Rapid Prototyping to Direct Digital Manufacturing*. 2010, New York: Springer.
10. Hopkinson, Neil, Hague, Richard J. M., and Dickens, Phillip, *Rapid Manufacturing: An Industrial Revolution for the Digital Age*. 2006, New York: John Wiley & Sons.
11. Sachs, Emmanuel, Cima, Michael, Williams, P., Brancazio, D., and Cornie, James, Three dimensional printing: rapid tooling and prototypes directly from a CAD model. *Journal of Engineering for Industry*, 1992. **114**(4): p. 481–488.
12. Neff, Clayton, Elston, Edwin, Cabrera, Ansell, Rojas-Nastrucci, Eduardo, Roberts, Blake, Weiss, Steve, Nguyen, Quang, Anthony, Theodore, Keetil, Manoj, Middleton, James, and Schrand, Amanda, Proof of concept additively manufactured (AM) conformal antenna on multi-option fuze artillery (MOFA) for survivability in harsh environments. *Journal of Defense Research and Engineering*, 2023, **6**(4): p. 40–57.

13. Schrand, Amanda, Kolel-Veetil, Manoj, Elston, Edwin, Neff, Clayton, Ajayi, Tosin, and Xue, Cheryl, *Printable Materials for Additive Manufacturing in Harsh Earth and Space Environments*. 2021, Singapore: Jenny Stanford Publishing.

14. Neff, Clayton, Elston, Edwin, Burfeindt, Matthew, Crane, Nathan, and Schrand, Amanda, A fundamental study of printed ink resiliency for harsh mechanical and thermal environmental applications. *Additive Manufacturing*, 2018. **20**: p. 156–163.

15. Neff, Clayton, Elston, Edwin, and Schrand, Amanda, Interconnections for additively manufactured hybridized printed electronics in harsh environments. *Designs*, 2020. **4**: p. 14, https://doi.org/10.3390/designs4020014.

16. *Production Equipment and Digital Solutions for the Additive Age*. 2020; Available from: https://optomec.com/.

17. *NanoJet Aerosol Print Technology*. 2021; Available from: https://www.idsnm.com/.

18. *Changing the way the world manufactures*. 2023; Available from: https://www.nano-di.com/.

19. *Product Realization for a High-Tech World*. 2023; Available from: https://www.neo-tech.com/.

20. *High Quality 3D Printers Made in the USA*. 2023; Available from: https://www.hyrel3d.com/.

21. *Industrial Precision Microdispensing and Direct Digital Manufacturing Equipment and Solutions*. 2020; Available from: https://www.nscrypt.com/.

22. ASTM, *Standard Test Methods for Rating Adhesion by Tape Test*, in D3359-22. 2022.

23. ASTM, *F1842 – 15 Standard Test Method for Determining Ink or Coating Adhesion on Flexible Substrates for a Membrane Switch or Printed Electronic Device*. 2015.

24. Gardner, Paul. *Gardco Paint Adhesion Test Kit*. 2017; Available from: https://www.gardco.com/pages/adhesion/PATkit.cfm.

25. *Flexible Electronics Standards*. 2023; Available from: https://www.semi.org/en/communities/flextech/FHE_standards.

26. Ashcroft, I. A. and Derby, B., Adhesion testing of glass-ceramic thick films on metal substrates. *Journal of Materials Science*, 1993. **28**: p. 2989–2998.

27. Hitch, T. T., *Adhesion Measurements on Thick-Film Conductors, in Adhesion Measurement of Thin Films, Thick Films, and Bulk Coatings, ASTM STP 640*, K. L. Mittal, Editor. 1978, New York: ASTM. p. 211–232.

28. Mittal, K. L., Adhesion measurement of thin films. *Electrocomponent Science and Technology*, 1976. **3**: p. 21–42.

29. Mittal, K. L. and Pizzi, A., *Adhesion Promotion Techniques*. 1999, New York: Marcel, Dekker, Inc.

30. Schirmer, Julian, Roudenko, Jewgeni, Reichenberger, Marcus, Neermann, Simone, and Franke, Jorg, Adhesion measurements for printed electronics: a novel approach to cross-cut-testing, *LOPEC –International Conference for the Printed Electronic Industry*. March 2018, Poster.

31. Volinsky, Alex A., Moody, Neville R., and Gerberich, William W., Interfacial toughness measurements for thin films on substrates. *Acta Materialia*, 2002. **50**: p. 441–460.

32. Neff, Clayton, *Analysis of Printed Electronic Adhesion, Electrical, Mechanical, and Thermal Performance for Resilient Hybrid Electronics*. 2018, University of South Florida: USF Tampa Graduate Theses and Dissertations.

33. Department of Defense, *MIL STD 2002.3 F Mechanical Shock*. 1996.

34. Roberson, David A., Wicker, Ryan B., Murr, Lawrence E., Church, Ken, and MacDonald, Eric, Microstructural and process characterization of conductive traces printed from Ag particulate inks. *Materials*, 2011. **4**: p. 963–979.

35. Mejias-Morillo, Carlos R., Gbaguidi, Audrey, Kim, Dae Won, Namilae, Sirish, and Rojas-Nastrucci, Eduardo A. UHF RFID-based additively manufactured passive wireless sensor for detecting micrometeoroid and orbital debris impacts. *2019 IEEE International Conference on Wireless for Space and Extreme Environments (WiSEE)*, 2019.

36. Ramirez, Ramiro, Rojas-Nastrucci, Eduardo A., and Weller, Thomas, 3D tag with improved read range for UHF RFID applications using additive manufacturing. *2015 IEEE 16th Annual Wireless and Microwave Technology Conference (WAMICON)*, 2014.

37. Ramirez, Ramiro, Rojas-Nastrucci, Eduardo A., and Weller, Thomas, UHF RFID tags for On-/Off-metal applications fabricated using additive manufacturing. *IEEE Antennas and Wireless Propagation Letters*, 2017. **16**: p. 1635–1638.

38. Rojas-Nastrucci, Eduardo A., Snider, Arthur D., and Weller, Thomas M., Propagation characteristics and modeling of meshed ground coplanar waveguide. *IEEE Transactions on Microwave Theory and Techniques*, 2016. **64**(11): p. 3460–3468.

39. Yu, Seng and Rojas-Nastrucci, Eduardo, Characterization of microdispensed dielectric materials for direct digital manufacturing using coplanar waveguides. *2019 IEEE 20th Wireless and Microwave Technology Conference (WAMICON)*, 2019. p. 1–3.

40. Rojas-Nastrucci, Eduardo A., Ramirez, Ramiro, and Weller, Thomas M. Direct digital manufacturing of mm-wave vertical interconnects. *2018 IEEE 19th Wireless and Microwave Technology Conference (WAMICON)*, 2018.

41. Adams, Jacob J., Duoss, Eric B., Malkowski, Thomas F., Motala, Michael J., Ahn, Bok Yeop, Nuzzo, Ralph G., Bernhard, Jennifer T., and Lewis, Jennifer A., Conformal printing of electrically small antennas on three-dimensional surfaces. *Advanced Materials*, 2011. **23**: p. 1335–1340.

42. Ketterl, T. P., Vega, Y., Arnal, N. C., Stratton, J. W. I., Rojas-Nastrucci, E. A., Cordoba-Erazo, M. F., Abdin, M. M., Perkowski, C. W., Deffenbaugh, P. I., Church, K. H., and Weller, T. M., A 2.45 GHz phased array antenna unit cell fabricated using 3-D multi-layer direct digital manufacturing. *Microwave Theory and Techniques, IEEE Transactions on Antennas and Propagation*, 2015. **63**(12): p. 4382–4394.

43. Mejias-Morillo, Carlos R. and Rojas-Nastrucci, Eduardo A. *Z-Meandering Miniaturized Patch Antenna Using Additive Manufacturing*. 2020 IEEE Radio and Wireless Symposium (RWS), 2020.

44. O'Brien, Jonathan M., Grandfield, John E., Mumcu, Gokhan, and Weller, Thomas W., Miniaturization of a spiral antenna using periodic Z-plane meandering. *IEEE Transactions of Antennas Propagation*, 2015. **63**(4): p. 1843–1848.

45. Rojas-Nastrucci, Eduardo A., *High performance digitally manufactured microwave and millimeter-wave circuits and antennas, Electrical Engineering.* 2017, University of South Florida: Scholar Commons.

46. Abdin, Mohamed, Johnson, W. Joel D., Wang, Jing, and Weller, Thomas, W-band MMIC chip assembly using laser-enhanced direct print additive manufacturing. *IEEE Transactions on Microwave Theory and Techniques*, 2021. **69**(12): p. 5381–5392.

47. Ramirez, Ramiro, Rojas-Nastrucci, Eduardo A., and Weller, Thomas, Laser-assisted additive manufacturing of mm-wave lumped passive elements. *IEEE Transactions on Microwave Theory and Techniques*, 2018. p. 1–10.

48. Rojas-Nastrucci, Eduardo A., Ramirez, Ramiro, Hawatmeh, Derar, Lan, Di, Wang, Jing, and Weller, Thomas. Laser enhanced direct print additive manufacturing for mm-wave components and packaging. *2017 International Conference on Electromagnetics in Advanced Applications (ICEAA)*, 2017.

49. Rojas-Nastrucci, Eduardo A., Tsang, Harvey, Deffenbaugh, Paul I., Ramirez, Ramiro A., Hawatmeh, Derar, Ross, Anthony, Church, Kenneth, and Weller, Thomas M., Characterization and modeling of K-band coplanar waveguides digitally manufactured using pulsed picosecond laser machining of thick-film conductive paste. *IEEE Transaction on Microwave Theory and Techniques*, 2017.

50. Jang, Keon-Soo, Eom, Yong-Sung, Choi, Kwang-Seong, and Bae, Hyun-Cheol, Crosslinkable deoxidizing hybrid adhesive of epoxy-diacid for electrical interconnections in semiconductor packaging. *Polymer International*, 2018. **67**(9): p. 1241–1247.

51. Sancaktar, Erol and Bai, Lan, Electrically conductive epoxy adhesives. *Polymers*, 2011. **3**(1): p. 427–466.
52. Bai, Guofeng, *Low-Temperature Sintering of Nanoscale Silver Paste for Semiconductor Device Interconnection*. 2005, New York: Virgina Tech.
53. Berry, David, Jiang, L., Mei, Yunhui, Luo, S., Ngo, Khai, and Lu, G. Q. Packaging of high-temperature planar power modules interconnected by low-temperature sintering of nanosilver paste. *2014 International Conference on Electronics Packaging (ICEP)*, 2014.
54. Lei, Thomas G., Calata, Jesus, Luo, Shu Fang, Lu, Guo Quan, and Chen, Xu, Low-temperature sintering of nanoscale silver paste for large-area joints in power electronics modules. *Key Engineering Materials*, 2007. **353–358**: p. 2948–2953.
55. Lu, Guo Quan, Calata, Jesus N., and Lei, Thomas G. Low-temperature sintering of nanoscale silver paste for power chip attachment. *5th International Conference on Integrated Power Electronics Systems*, 2008.
56. Andersson, Henrik A., Manuilskiy, Anatoliy, Haller, Stefan, Hummelgård, Magnus, Sidén, Johan, Hummelgård, Christine, Olin, Håkan, and Nilsson, Hans-Erik, Assembling surface mounted components on ink-jet printed double sided paper circuit board. *Nanotechnology*, 2014. **25**(9): p. 094002.
57. Arrese, J., Vescio, Giovanni, Xuriguera, Elena, Medina-Rodriguez, Beatriz, Cornet, A., and Cirera, Albert, Flexible hybrid circuit fully inkjet-printed: surface mount devices assembled by silver nanoparticles-based inkjet ink. *Journal of Applied Physics*, 2017. **121**(10): p. 104904.
58. Bagrets, Nadezda, Barth, Christian, and Weiss, Klaus-Peter, Low temperature thermal and thermo-mechanical properties of soft solders for superconducting applications. *IEEE Transactions on Applied Superconductivity*, 2014. **24**(3): p. 1–3.
59. Kim, Keun-Soo, Imanishi, Takayuki, Suganuma, Katsuaki, Ueshima, Mimoru, and Kato, Rikiya, Properties of low temperature Sn-Ag-Bi-In solder systems. *Microelectronics Reliability*, 2007. **47**(7): p. 1113–1119.
60. Li, Qin, Lei, Yongping, Lin, Jian, and Yang, Sai. Design and properties of Sn-Bi-In low-temperature solders. *2015 16th International Conference on Electronic Packaging Technology (ICEPT)*, 2015.
61. Mei, Zequn, Hua, Fay, Glazer, J., and Chung, C. Key. Low temperature soldering. *Twenty First IEEE/CPMT International Electronics Manufacturing Technology Symposium Proceedings 1997 IEMT Symposium*, 1997.
62. Ren, Guang, Wilding, Ian J., and Collins, Maurice N., Alloying influences on low melt temperature SnZn and SnBi solder alloys for electronic interconnections. *Journal of Alloys and Compounds*, 2016. **665**: p. 251–260.
63. Sahasrabudhe, Shubhada H., Mokler, Scott, Renavikar, Mukul, Sane, Sandeep, Byrd, Kevin, Brigham, Eric, Jin, Owen, Goonetilleke, Pubudu, Badwe, Nilesh, and Parupalli, Satish. Low temperature solder – a breakthrough technology for surface mounted devices. *2018 IEEE 68th Electronic Components and Technology Conference (ECTC)*, 2018.
64. Wu, Albert T., Chen, Chih-Hao, Huang, Jyun-Jhe, Chiang, Jeng-Yu, and Wang, Chang-Meng. Development of low temperature solder alloys for advanced electronic packaging: assessment of In-Bi alloys on Cu substrates. *2018 International Conference on Electronics Packaging and iMAPS All Asia Conference (ICEP-IAAC)*, 2018.
65. Yang, Chih-han, Zhou, Shiqi, Lin, Shih-kang, and Nishikawa, Hiroshi, A computational thermodynamics-assisted development of Sn-Bi-In-Ga quaternary alloys as low-temperature Pb-free solders. *Materials*, 2019. **12**(4): p. 631.

66. Liu, Pilin L. and Shang, Jianku K., Interfacial embrittlement by bismuth segregation in copper/tin-bismuth Pb-free solder interconnect. *Journal of Materials Research*, 2011. **16**(6): p. 1651–1659.
67. Un-Byoung, Kang and Young-Ho, Kim. Electrical characteristics of fine pitch flip chip solder joints fabricated using low temperature solders. *2004 Proceedings. 54th Electronic Components and Technology Conference (IEEE Cat. No.04CH37546)*, 2004.
68. Bai, H., Xu, F., Sha, W., Chen, D., Yan, J., and Gan, Y., Phase structure, microstructure and properties of Sn-Bi-Ag solder alloy in ternary system. *Rare Metal*, 2019. **43**(1): p. 44–51.
69. Made, Riko I., Gan, Chee Lip, Yan, Li Ling, Yu, Aibin, Yoon, Seung Wook, Lau, John H., and Lee, Chengkuo, Study of low-temperature thermocompression bonding in Ag-In solder for packaging applications. *Journal of Electronic Materials*, 2008. **38**(2): p. 365.
70. Mei, Zequn, Vander Plas, Hubert, Gleason, and Baker, J.. Low-temperature solders. *Proceedings of the Electronic Materials and Processing Symposium*, 1994.
71. Youngil, L., Jun-rak, C., Jong, L. Kwi, Nathan, E. S., and Donghoon, K., Large-scale synthesis of copper nanoparticles by chemically controlled reduction for applications of inkjet-printed electronics. *Nanotechnology*, 2008. **19**(41): p. 415604.
72. Basaran, C., Tang, H., Dishongh, T., and Searls, D., Computer Simulations of Solder Joint Reliability Tests Electronic Packaging Laboratory, Advanced Packaging, 2001. **1**(Review article #27): p. 17–22.
73. Johnson, R. Wayne, Extreme temperature packaging: challenges and opportunities. *Proceedings SPIE 9836, Micro- and Nanotechnology Sensors, Systems, and Applications VIII*, 98360L, 2016.
74. Johnson, R. Wayne, Zheng, Ping, Wiggins, Alberez, Rubin, Seymour, and Peltz, Leora, High Temperature Electronics Packaging. 2007. https://www.researchgate.net/publication/228998676_High_Temperature_Electronics_Packaging
75. Sung, Yong-Gue, Myung, Woo-Ram, Jeong, Haksan, Ko, Min Kwan, Moon, Jeonghoon, and Jung, Seung-Boo, Mechanical reliability of the epoxy Sn-58wt.%Bi solder joints with different surface finishes under thermal shock. *Journal of Electronic Materials*, 2018. **47**: p. 4165–4169
76. Ribas, Morgana, Chegudi, Sujatha, Kumar, Anil, Pandher, Ranjit, Raut, Rahul, Mukherjee, Sutapa, Sarkar, Siuli, and Singh, Bawa. Thermal and mechanical reliability of low-temperature solder alloys for handheld devices. *2014 IEEE 16th Electronics Packaging Technology Conference (EPTC)*, 2014.
77. Rahman, Taibur, Renaud, Luke, Heo, Deukhyoun, Renn, Michael, and Panat, Rahul, Aerosol based direct-write micro-additive fabrication method for sub-mm 3D metal-dielectric structures. *Journal of Micromechanics and Microengineering*, 2015. **25**: p. 107002.
78. Cai, Fan, Pavlidis, Spyridon, Papapolymerou, John, Chang, Yung-hang, Wang, Kan, Zhang, Chuck, and Wang, Ben, Aerosol jet printing for 3-D multilayer passive microwave circuitry. *2014 44th European Microwave Conference*, 2014: p. 512–515.
79. Mahajan, Ankit, Frisbie, C. Daniel, and Francis, Lorraine F., Optimization of aerosol jet printing for high-resolution, high-aspect ratio silver lines. *ACS Applied Materials & Interfaces*, 2013. **5**(11): p. 4856–4864.
80. Cai, Fan, Chang, Yung-hang, Wang, Kan, Khan, Wasif Tanveer, Pavlidis, Spyridon, and Papapolymerou, John, High resolution aerosol jet printing of D- band printed transmission lines on flexible LCP substrate. *2014 IEEE MTT-S International Microwave Symposium (IMS2014)*, 2014: p. 1–3.

81. Wadhwa, Arjun, Cormier, Denis, and Williams, Scott, Improving run-time stability with aerosol jet printing using a solvent add-back bubbler. *The Journal of Print and Media Technology Research*, 2016. **5**: p. 207–214.

82. Secor, Ethan B., Principles of aerosol jet printing. *Flexible and Printed Electronics*, 2018. **3**(3): p. 035002.

83. Smith, Michael, Choi, Yeon Sik, Boughey, Chess, and Kar-Narayan, Sohini, Controlling and assessing the quality of aerosol jet printed features for large area and flexible electronics. *Flexible and Printed Electronics*, 2017. **2**(1): p. 015004.

84. Wilkinson, Nathan, Smith, M. A. A., Kay, Robert W., and Harris, Russel, A review of aerosol jet printing—a non-traditional hybrid process for micro-manufacturing. *The International Journal of Advanced Manufacturing Technology*, 2019. **105**(11): p. 4599–4619.

85. Chen, Guang, Gu, Yuan, Tsang, Harvey, Hines, Daniel R., and Das, Siddhartha, The effect of droplet sizes on overspray in aerosol-jet printing. *Advanced Engineering Materials*, 2018. **20**(8): p. 1701084.

86. *About NextFlex: Building the Next Big Thing in Flexible Hybrid Electronics.* 2018; Available from: https://www.nextflex.us/about/.

87. www.IDTechEx.com

3 Additively Manufactured Antennas for Aerospace Harsh Environments

Eduardo A. Rojas-Nastrucci
Embry-Riddle Aeronautical University

3.1 INTRODUCTION

Additive manufacturing (AM) technologies have inspired extensive innovations in antennas and wireless devices. AM processes can fabricate antennas that not only meet the performance of their traditionally manufactured counterparts but, in some cases, have enabled new antennas with superior performance that take advantage of (1) new materials, (2) advanced 3D geometries, and (3) the ability to create materials with engineered graded properties. Furthermore, AM allows for the direct translation from the virtual design and simulation world to a 3D part, enabling for a genuine cyber-physical connection. As AM processes advance to achieve better feature sizes, materials properties, and reliability, the value of AM antennas is increasingly being embraced for aerospace applications.

The state of the art of AM for antennas has, for example, highly complex antennas that would not be possible to manufacture as a single piece with traditional machining [1,2] that offer size and weight advantages. Novel AM processes are allowing the fabrication of high-performance antenna materials that exhibit great compatibility for harsh environments with high temperatures (>500°C) and high g in hypersonic applications [3], as well as space missions with extreme temperature changes, gamma radiation, vacuum, and vehicle launch vibrations [4]. Figure 3.1 shows three examples of 3D antennas enabled by AM with different processes and applications. Figure 3.1a shows a picture of a complex Luneburg lens from the pioneering work presented in [1] fabricated with ceramic stereolithography and achieving a gain of 26 dB at 33 GHz. On the other hand, Figure 3.1b shows a z-directed meandering technique to accomplish an ultra-wideband 0.8–3 GHz spiral antenna [5] that achieves a footprint reduction of 76.7% compared to a traditional flat design. Another example of the geometries that AM has enabled is shown in Figure 3.1c, which presents a three-dimensional dielectric Ka-band reflect array with a 30% 3-dB gain fractional bandwidth, surpassing the performance of similarly sized traditional microstrip versions [6]. The three samples of 3D antennas presented in Figure 3.1 represent a microcosm of a rapid increase of innovative geometries generated by the engineering and scientific community, enabled by AM.

DOI: 10.1201/9781003138945-3

31

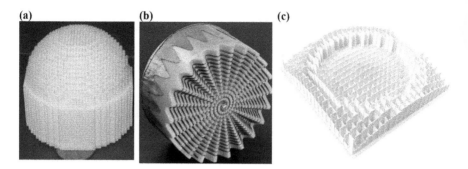

FIGURE 3.1 Examples of 3D antennas using AM. (a) Ceramic Luneburg lens [1]. © IEEE 2007. (b) 3D Z-plane meandered spiral antenna [5]. © IEEE 2015. (c) 3D broadband dielectric reflect array antennas [6]. © IEEE 2022.

AM of antennas is gaining terrain into commercial applications where it brings a great value-added, i.e., in two general scenarios: 1) when the complexity of the design is prohibitively expensive or impossible to manufacture with traditional techniques, or 2) when the low quantities of specialized high-performance design are not compatible with the high-volume manufacturing industries. However, as the industry strives for AM processes with higher throughput, lower cost, better accuracy, and feature sizes, AM is envisioned to be adopted and embraced by a large section of the antenna manufacturing industry over the next decades. Furthermore, AM offers a highly strategic mission value for defense applications where manufacturing can be performed ad hoc and in the field, as well as for space exploration missions where it is essential to fabricate electronics on-site as it may take years to obtain hardware replacements.

This chapter intends to offer the reader an overview of the AM processes used for antennas. More detailed exploration is provided for AM antennas used for harsh environments such as ones with high temperatures, high-g impacts, vibrations, vacuum, and significant temperature changes, to name a few, as well as antennas for aerospace applications such as ones for hypersonic bodies and spacecraft. The goal is to give the reader a general understanding of the state-of-the-art AM for aerospace applications.

3.2 ADDITIVE MANUFACTURING PROCESSES FOR ANTENNAS

Antennas are often one of the most challenging components to test an AM process because they are susceptible to dimensional accuracy, material properties, and surface finish. Furthermore, antennas often are embodied with complex geometries, which are the perfect case to showcase the power of AM. The research community has explored a wide range of AM technologies to advance state-of-the-art wireless systems for applications that range from consumer electronics to defense applications. This section provides an overview of AM processes that allow the successful fabrication of dielectric and electrically conductive layers to realize antennas and their related radiofrequency (RF) components.

Direct-print additive manufacturing (DPAM): DPAM is a versatile technology that allows the micro-dispensing mechanism to control the 3D deposition of electrical conductors and dielectrics. DPAM systems can deposit thick films with thickness typically within 10–100 μm [7,8] of pastes with viscosity in the range of 1–1M cps, and control down to deposition of 100 pl. DPAM systems manufactured by nScrypt are often equipped with fused filament fabrication (FFF) heads to achieve deposition of dielectric filaments that can be loaded with ceramics to achieve a range of dielectric constants of 2.1–8.72 [9–11]. Silver conductive inks such as Dupont's CB028 [12,13] and Nova Centrix's HPS-FG77 [14,15] can be used with the DPAM systems to achieve feature sizes down to 50 μm [16]. Other inks, such as the XTPL CL85, have been optimized to achieve down to the 1–10 μm line widths range using the micro-dispensing technology [17]. The DPAM systems have been successfully used to manufacture 3D and conformal antennas, such as the tripolar antenna described in [18], which offer advantages for communication over clutter channels. It has also been used to manufacture stacked structures with passive distributed elements and embedded lumped passive and active components to achieve a 2.45 GHz phased array antenna unit cell [7]. The DPAM process can also print conformal thick films of dielectrics [19] and multilayer dielectric-conductor stacks [14].

DPAM systems have been enhanced with picosecond to femtosecond lasers (LE-DPAM) to reduce the minimum achievable features below 10 μm over silver paste layers that are 20 μm thick [16,20]. LE-DPAM also allows for precise laser machining of dielectrics to achieve vertical interconnects in multilayer structures [14,21] and component structural embedding. The laser enhancement has enabled the use of this technology to achieve components that operate at mm-wave frequencies, such as the monolithic microwave-integrated circuit (MMIC) interconnects shown in [22] that work up to 110 GHz.

Aerosol jet printing (AJP): The material deposition with this technology is based on precise control of an atomized steam of materials with a viscosity typically below 1,000 cp [23]. When using conductive materials optimized for the AJP process on highly polished and compatible substrates, it is possible to achieve electrically conductive traces and dielectric layers with feature sizes down to 10 μm [24] for a layer thickness that is below 1 μm per pass. Microwave circuits and antennas can also be manufactured using aerosol jet printing—the selective deposition of nanoparticle ink droplets onto a substrate—achieving low-loss dielectrics and conductors [25,26].

Nanoparticle jetting (NPJ) is another technology that allows for AM of high-temperature dielectrics [3,27] with an ink jetting resolution down to 20 μm and a layer thickness of 10 μm [28]. Yttria-stabilized zirconia manufactured using the NPJ process has been characterized for electrical properties in [29], showing a bulk permittivity of 23 and a loss tangent of 0.0013 in the range of 1.68–3.99 GHz.

Laser powder bed fusion (LPBF), also trademarked as selective laser sintering (SLS), selective laser melting (SLM), and laser beam melting (LBM), is a process that utilizes lasers as a source of selective heating that is focused on a powder bed to create a part layer by layer [30]. The LBPF can produce nearly 100% dense parts using metal alloys with roughness in the order of 12.89 μm arithmetical mean height (Sa) [31], which can be improved by additional metal plating once the part is printed. Another technology used to print solid metal parts is the binder jetting (BJ) process, where a layer of metallic or ceramic powder is spread and an inkjet selectively deposits a liquid binder agent in areas of the layer where it is desired to obtain a solid part [32]. After the printing process and sintering, it produces a solid matrix of metal powder particles fused together [32]; this matrix can be infiltrated with a second metal with a lower melting point than the one from the initial powder to achieve nearly 100% dense composite material [33,34].

Stereolithography (SL, SLA), also known as vat polymerization, is another process that has been demonstrated to produce high-performance antennas and rectangular waveguide components when plated with a metal layer [35]. In the SL process, a photo-resin is exposed to a laser beam layer by layer to solidify it and form a 3D object with a typical layer resolution of around 25 μm. After copper plating, SL parts can achieve a surface roughness (arithmetic average) of 0.93 μm [35]. SL typically produces parts with better surface roughness than the FFF process, which, when plated with metals such as copper, can achieve low-loss microwave frequency structures [36]. A variation of the SL process can achieve ceramic parts by replacing the resin with a photoreactive ceramic suspension that is then sintered to achieve a dense ceramic part [1].

3.3 ADVANCED ANTENNAS FOR HIGH-TEMPERATURE AND HYPERSONIC APPLICATIONS

The sub-set of AM materials and manufacturing processes that can endure temperatures greater than 300°C is significantly reduced since most polymer-based dielectrics and resin-based conductive inks and past cannot survive such an extreme environment. On the other hand, the interest in antennas for high-temperature and hypersonic environments has grown as AM enables the manufacturing of conformal antennas with high-temperature materials. This section offers an overview of the hypersonic antenna environment, advancements in high-temperature antenna AM technologies, and antenna design aspects related to the traditional cylindrical or conical geometries inherent in hypersonic vehicles.

3.3.1 THE HYPERSONIC ANTENNA ENVIRONMENT

Aircraft, spacecraft, and munitions heavily rely on wireless communications and sensing for their operation while often traveling at a ground speed greater than Mach 5 through the atmosphere. These hypersonic speeds generate a high-temperature, high-pressure, and ablating environment for which antennas are required to operate and survive. The environmental conditions heavily depend on the location of the antenna along the body of the vehicle, as well as the altitude, speed, angle of attack, and materials properties, and it can be modeled using computational fluid dynamics (CFD) techniques [37,38]. Experimental data from the North American X-15 aircraft has extensively been used to validate CFD models, and it shows that for speeds up to Mach 4 temperatures on the leading edge of the wings can reach temperatures >300°C [37]. More recent CFD simulations of a realistic antenna radome with a circular aperture of 0.2 m, height of 0.92 m, and traveling at speeds up to Mach 15.6 predict a radome surface temperature of up to 2,100°C during flight [38]. Figure 3.2

(a)   (b)

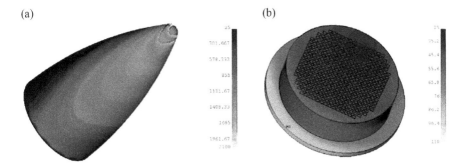

FIGURE 3.2   Surface temperature profile for (a) radome and (b) phased array antenna inside of the radome for $t = 300$ seconds of flight [38]. © IEEE 2020.

shows the simulated temperature profile of the surface of a radome (Figure 3.2a) and a phased array antenna (Figure 3.2b) inside of the radome at $t = 300$ s for a flight described in [38]. The National Aeronautics and Space Administration (NASA) is equipped with facilities that can emulate environments of speeds up to Mach 12 and temperatures up to 1,727°C, allowing for the development of hypersonic vehicle technologies and confirming they can survive the harsh hypersonic environment [39].

Antennas for hypersonic environments must be designed with materials that can withstand high temperatures and pressures during the flight but also the ablation of the structure. In [38], a comprehensive study presents predicted temperature and ablation on a hypersonic phased array antenna radome and gives the theoretical basis for compensating the effects of radome ablation to preserve the antenna performance. The high-temperature environment will inevitably lead to the thermal expansion of airborne radomes. Variable thickness radome (VTR) techniques have been proposed in the literature to compensate for thermal expansion effects and preserve the antenna performance during flight [40]. Another relevant hypersonic antenna design aspect is the effect of the plasma sheath. A plasma sheath is a lossy and inhomogeneous layer of electrons, ions, and other neutral particles that forms around the structure at hypersonic speeds [41]. The effects of the plasma sheath on antennas can be modeled with layers of relative dielectric permittivity (ε_r) and loss tangents ($\tan \delta$) that are a function of environmental parameters [41], with studies showing that antenna radiation patterns can greatly degrade under hypersonic conditions [42]. The antenna effects of the plasma sheath show a strong dependency on the operating frequency, with models showing that millimeter-wave frequencies may be preferred for lower losses and reduced non-linear effects [43,44]

3.3.2   High-Temperature Antennas and Sensors

The main factor constraining AM for hypersonics is the maximum operational temperature of materials. Hence, antenna engineers seek to find materials that can withstand the temperature requirements while having the highest electric conductivity possible (σ) for electric conductors and the lowest loss tangent ($\tan \delta$) for the dielectrics. Another important factor for dielectrics is the relative permittivity's stability as a function of temperature [45]. For example, planar passive wireless sensors based on a patch antenna, fabricated by traditional means using alumina and platinum, have been demonstrated to survive up to 1,050°C [46]. Other examples of 3D high-temperature wireless components include an antenna integrated with a resonant cavity to form an evanescent-mode resonator-based pressure sensor that can be fabricated using polymer-derived ceramic (PDC) [47] based on silicoaluminum carbonitride (SiAlCN) and platinum; and a picosecond laser machined [48] alumina structure with platinum metallization that can survive up to 1,700°C.

Among the most common additively manufactured high-temperature dielectrics for antennas are Alumina and Zirconia [3]. A highly efficient dielectric spiral antenna [49] and an all-dielectric helical antenna [50] using zirconia are manufactured using SL. Dielectric rod antennas (DRA) have also been demonstrated using zirconia fabricated using the NPJ process [27,29]. Dielectric antennas and lenses with complex 3D geometries enabled by AM, such as the gradient index lens demonstrated in [51,52], show that NPJ can create a volumetric distribution of electric permittivities by controlling the volumetric fraction of ceramic across the geometry.

Silver-particle-based pastes [7,14,15,53] and inks [23,24] are among the most widely used conductors for additively manufactured antennas. DPAM is also able to deposit other high-temperature materials, such as platinum inks [3]. It has been demonstrated that platinum inks deposited using the DPAM process on NPJ Yttria-stabilized zirconia can be used to fabricate antennas that survive temperatures above 600°C [3]. The two-layer back-fed antenna design presented in [3] allows for having the RF electronics inside of a hypothetical radome, while having the radiating elements on the outer surface (Figure 3.3a). Figure 3.3b shows the top and bottom pictures of the additively manufactured micro-dispensed antenna with platinum (Pt) ink as the black color after drying before sintering. The antenna design in [3] shows a 2.5 dBi gain at 4.1 GHz, with a 160 MHz return loss bandwidth (Figure 3.3c). Characterization of the electromagnetic properties, up to 6 GHz, of the AM platinum and Yttria-stabilized zirconia is performed using coplanar waveguides (CPW) under thermal fatigue testing, showing no degradation after 9 cycles of heating up to 600°C, maintaining this temperature for 30 minutes, and cooling down [3].

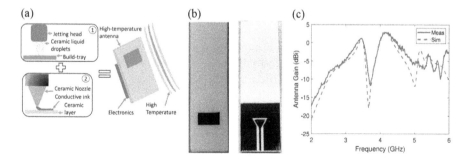

FIGURE 3.3   AM high-temperature antenna. (a) AM process description, (b) picture of manufactured antenna using silver Pt ink and a Yttria-stabilized zirconia dielectric, and (c) measured peak antenna gain vs. frequency [3]. © IEEE 2022.

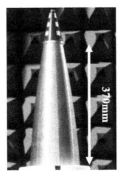
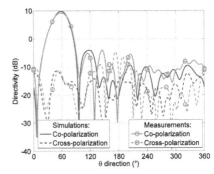

FIGURE 3.4   Conformal patch antenna array on a conical arrangement. (a) Isometric view of the manufactured antenna array, (b) conical array mounted on a projectile shell, and (c) measured antenna directivity pattern [61]. © IEEE 2017.

The DPAM technology can be used to micro-dispense thick films of a low-temperature cofired ceramic (LTCC) paste, which is sintered to fabricate a high-density ceramic substrate that can survive high temperatures [54]. A planar miniaturized Ka-band antenna has been achieved in [54], combining micro-dispensing of a ceramic paste, sintering, dispensing of silver and platinum (Pt) pastes, and pulsed laser machining, with impedance resonances in the range of 32–34 GHz. The process in [54] achieves CPW structures with losses lower than 1 dB/mm at 40 GHz and an effective relative electric permittivity of around 20.

3.3.3 Design of Antennas for Conical and Cylindrical Geometries

Antennas for hypersonic applications are often either located inside of a radome or on a surface that can be cylindrical or conical in nature. Classic theoretical studies in the literature cover the effects of patch antennas with cylindrical curvatures [55–57]. Other antenna geometries, such as the spiral antenna [58,59] and rectangular patches [58] that conform to a cone, show ways that the antenna can be taken from the inside of the radome to its inner or outer surface. For applications where beam scanning is desired, antenna-phased arrays are shown as a viable option on conical shapes [60,61] and cylindrical sections [62,63]. In [61], a conical 12-element phased array (Figure 3.4a) is designed to be mounted on a 155 mm projectile shell (Figure 3.4b) with an adjustable pointing direction range of 30°–110° in the elevation plane. Figure 3.4c shows the simulated and measured radiation pattern for a pointing angle of 60° at an elevation plane of $\varphi = 0°$ [61]. The antenna array shown in Figure 3.4 is reported to have survived a launch environment with an acceleration of up to 16,000g

3.4 ADDITIVELY MANUFACTURED ANTENNAS FOR SPACE MISSIONS

The space environment offers an extensive set of challenges to antennas that heavily vary by the specifics of each mission. For example, for low Earth orbit (LEO) missions, antennas must withstand the effects of atomic oxygen [64], which strongly interacts with polymers, carbon, and many metals. Vacuum is another common condition in space that antenna engineers typically do not consider for commercial applications, but material outgassing can threaten missions that require optical sensors. Extreme temperature changes are yet another condition often found in space missions; the Moon surface temperature can be modeled using expressions from [65] and can be in the approximate range of −200°C to 87°C, for 35° Latitude, during a lunar day [66]; and hardware designed for the Venusian surface must withstand almost constant temperature of 460°C, a pressure of ~90 bar, and caustic chemical environment [67]. Another important space hardware hazard is vibrations during launch; tests for acceleration spectral density survival requirements along with the temperature, pressure, and other space-related environmental aspects are well defined in [68]. An extensive material selection for different space environments is presented in [64].

Antennas for traditional large satellites include reflector antennas due to their high efficiency and the ability to deploy large meshed apertures in space to achieve high antenna gains and multiple beams [69]. Another standard antenna geometry for satellites is metallic feed arrays that allow for tailoring an antenna gain pattern and its corresponding coverage on earth, for instance, continental coverage for geostationary missions. In contrast, small satellites are heavily constrained in size, weight, and power, offering a more significant challenge for antenna designers. Among popular selections for CubeSat antennas are patch antenna arrays, reflectarrays [70], and metasurface antennas [71].

The power of AM is embraced for highly complex geometries that push the limits of traditional computer numerically controlled (CNC) machining. An example of such geometry is all-metal metasurface (MTS) antennas with hundreds or even thousands of uniquely shaped and positioned elements to emulate a target current distribution at the antenna aperture. Figure 3.5a shows the top surface of a Ka-band

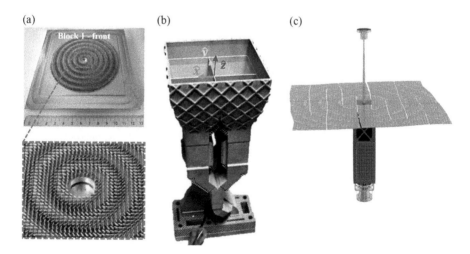

(a)   (b)   (c)

FIGURE 3.5   (a) All-metal AM Ka-band meta surface antenna [4], © IEEE 2018. (b) AM Ku-band horn antenna with integrated with the feed network [2]. © IEEE 2022. (c) Origami-based deployable CubeSat X-band antenna design [72]. © IEEE 2020.

MTS [4] manufactured by LPBF with the AlSi10Mg aluminum-based alloy using a ProX DMP 320 and a resulting roughness of around 5 μm; the resulting antenna shows a peak right-hand circularly polarized (RHCP) gain of 26.1 dBi and a 40% aperture efficiency.

All-metal antennas are often preferred for space environments due to their natural endurance for vibrations, extreme temperatures, and ionizing radiation. In [2], AM is embraced to realize a highly complex feed network—a 4-way dual-polarized orthomode waveguide power divider—integrated with a horn antenna to achieve an antenna aperture efficiency >90% for Ku-band frequencies, all printed in a single part (Figure 3.5b). The quad-furcated profiled horn in [2] is made using AlSi10Mg by the LPBF process with tolerances of ±50 μm. Another example of an intricated antenna feed network component is the combination of three classic waveguide elements, namely, a twist, a bend, and a filter, in one piece inspired by AM [73]. The demonstration of such a combination in one component for the WR51 waveguide standard is shown in [73], where the geometry fabricated using the AlSi10Mg LPBF process would be impossible to CNC machine as a single part. BJ is an AM technology that can achieve high-density all-metal antenna parts [32–34]. In [33,74], the concept of meshed rectangular waveguides is introduced, and prototypes are fabricated using BJ, achieving an alloy of stainless steel and copper—meshed waveguides save weight with minimal effect on the performance of the components when compared with traditional structures. BJ can manufacture a meshed reflector antenna design with an integrated feed horn and filter, achieving a peak gain of 22.66 dBi at 32 GHz [32].

Reflectarray antennas can be seen as an engineered array of electromagnetic elements that produce a desired wave reflection phase to achieve a particular radiation characteristic while illuminated by an antenna feed. Reflectarrays offer the freedom of attaining flat aperture geometries that can be folded during deployment and deployed in space [70]. The Mars Cube One (MarCO) CubeSat mission successfully demonstrated an X-band reflect array antenna with 29.2 dBi RHCP gain that orbited Mars. Efforts to achieve multilayer reflect array antennas are described in [14] where a two-layer reflect array unit cell is fabricated using the DPAM process and characterized for K-band frequencies. Figure 3.5c shows the geometry of an origami-based X-band reflect array antenna [72] designed for a 3U CubeSat with a prototype showing a peak gain of 27.2 dBi for a deployed area of 0.41 m² from a stowed volume of 8.3×10^{-4} m². Among the critical factors for a successful AM space reflect array antenna are using dielectrics and conductive inks that meet the outgassing requirements and can withstand temperature fluctuations with minimal material degradation and deformation for the mission's duration.

A study in [75] shows that copper-plated AM microwave components manufactured by LPBF and qualified for space environments. In [75], a microwave low-pass filter design is adapted for the AM process and tested for the temperature range from −180°C to 200°C. Another all-metal AM process, the Nuvotronics PolyStrata, also called 3D copper additive manufacturing (3D-CAM), generates 3D geometries using a layer-by-layer metal deposition with a high accuracy compatible with requirements for millimeter-wave frequency applications [76,77]. A 94 GHz 2×1 antenna array unit cell using the Polystrata process is presented in [76] as a demonstration for a potential cloud profiling radar instrument.

3.5 CONCLUSIONS

The advancement of materials, the discovery of new geometries, and the development of new fabrication processes drive the art of antenna design and manufacturing. The challenges related to antenna engineering increase when seeking to survive harsh environments such as hypersonic flight or space. This chapter not only shows an overview of the harsh environmental factors to be considered in antenna design, but it also presents successful examples of AM antennas that can survive such conditions as high temperatures, high accelerations, high pressure, vacuum, and ionizing radiation. Furthermore, AM has enabled antenna engineers to explore the third dimension to achieve unique designs that offer performance, size, weight, or cost advantages. The engineering and scientific community will continue to see increased adoption of AM for antennas as the AM systems increase throughput and novel 3D antenna geometries are discovered. In the case of space exploration, AM is an indispensable tool to sustain the long-term settlement of humans on a distant planet, where a large variety of devices can be fabricated ad hoc with a single AM system.

REFERENCES

[1] K. F. Brakora, J. Halloran, and K. Sarabandi, "Design of 3-D monolithic MMW antennas using ceramic stereolithography," *IEEE Transactions on Antennas and Propagation*, vol. 55, no. 3, pp. 790–797, 2007, https://doi.org/10.1109/tap.2007.891855.
[2] C. Stoumpos, J. P. Fraysse, G. Goussetis, R. Sauleau, and H. Legay, "Quad-furcated profiled horn: the next generation highly efficient GEO antenna in additive manufacturing," *IEEE Open Journal of Antennas and Propagation*, vol. 3, pp. 69–82, 2022, https://doi.org/10.1109/OJAP.2021.3134833.
[3] C. R. Mejias-Morillo et al., "High-temperature additively manufactured C-band antennas using material jetting of zirconia and micro-dispending of platinum paste," *IEEE Open Journal of Antennas and Propagation*, vol. 3, pp. 1289–1301, 2022, https://doi.org/10.1109/OJAP.2022.3218798.
[4] D. González-Ovejero, N. Chahat, R. Sauleau, G. Chattopadhyay, S. Maci, and M. Ettorre, "Additive manufactured metal-only modulated metasurface antennas," *IEEE Transactions on Antennas and Propagation*, vol. 66, no. 11, pp. 6106–6114, 2018, https://doi.org/10.1109/TAP.2018.2869135.
[5] J. M. O'Brien, J. E. Grandfield, G. Mumcu, and T. M. Weller, "Miniaturization of a spiral antenna using periodic Z-plane meandering," *IEEE Transactions on Antennas and Propagation*, vol. 63, no. 4, pp. 1843–1848, 2015, https://doi.org/10.1109/TAP.2015.2394796.
[6] Q. Cheng et al., "Dual circularly polarized 3-D printed broadband dielectric reflectarray with a linearly polarized feed," *IEEE Transactions on Antennas and Propagation*, vol. 70, no. 7, pp. 5393–5403, 2022, https://doi.org/10.1109/TAP.2022.3142735.
[7] T. P. Ketterl et al., "A 2.45 GHz phased array antenna unit cell fabricated using 3-D multi-layer direct digital manufacturing," *IEEE Transactions on Microwave Theory and Techniques*, vol. 63, no. 12, pp. 4382–4394, 2015, https://doi.org/10.1109/TMTT.2015.2496180.
[8] R. Ramirez, E. Rojas-Nastrucci, and T. Weller, "UHF RFID tags for on/off-metal applications fabricated using additive manufacturing," *IEEE Antennas and Wireless Propagation Letters*, vol. PP, no. 99, pp. 1–1, 2017, https://doi.org/10.1109/lawp.2017.2658599.

[9] J. Castro, E. A. Rojas-Nastrucci, A. Ross, T. M. Weller, and J. Wang, "Fabrication, modeling, and application of ceramic-thermoplastic composites for fused deposition modeling of microwave components," *IEEE Transactions on Microwave Theory and Techniques*, vol. 65, no. 6, pp. 2073–2084, 2017, https://doi.org/10.1109/tmtt.2017.2655057.

[10] J. Castro, E. Rojas, T. Weller, and J. Wang, "High-permittivity and low-loss electromagnetic composites based on co-fired Ba0.55Sr0.45TiO3 or MgCaTiO2 microfillers for additive manufacturing and their application to 3-D printed K-band antennas," *Journal of Microelectronics and Electronic Packaging*, vol. 13, no. 3, pp. 102–112, 2016, https://doi.org/10.4071/imaps.509.

[11] F. Castles et al., "Microwave dielectric characterisation of 3D-printed BaTiO3/ABS polymer composites," *Scientific Reports,* vol. 6, p. 22714, 03/04/online 2016, https://doi.org/10.1038/srep22714.

[12] C. R. Mejias-Morillo, A. Gbaguidi, D. W. Kim, S. Namilae, and E. A. Rojas-Nastrucci, "UHF RFID-based additively manufactured passive wireless sensor for detecting micrometeoroid and orbital debris impacts," In *2019 IEEE International Conference on Wireless for Space and Extreme Environments (WiSEE),* 16–18 Oct. 2019 2019, pp. 41–47, https://doi.org/10.1109/WiSEE.2019.8920352.

[13] E. A. Rojas-Nastrucci, A. D. Snider, and T. M. Weller, "Propagation characteristics and modeling of meshed ground coplanar waveguide," *IEEE Transactions on Microwave Theory and Techniques,* vol. 64, no. 11, pp. 3460–3468, 2016, https://doi.org/10.1109/tmtt.2016.2606409.

[14] S. LeBlanc, K. Church, E. A. Rojas-Nastrucci, E. Martinez-de-Rioja, E. Carrasco, and J. A. Encinar, "Advanced manufacturing and characterization of mm-wave two-layer reflectarray cells," In *2022 IEEE 22nd Annual Wireless and Microwave Technology Conference (WAMICON),* 27–28 April 2022 2022, pp. 1–4, https://doi.org/10.1109/WAMICON53991.2022.9786075.

[15] S. LeBlanc, K. Church, and E. A. Rojas-Nastrucci, "Photonic curing of mm-wave coplanar waveguides for conductor loss enhancement," In *2021 IEEE 21st Annual Wireless and Microwave Technology Conference (WAMICON),* 28–29 April 2021, pp. 1–3, https://doi.org/10.1109/WAMICON47156.2021.9443580.

[16] E. A. Rojas-Nastrucci et al., "Characterization and modeling of K-band coplanar waveguides digitally manufactured using pulsed picosecond laser machining of thick-film conductive paste," In *IEEE Transactions on Microwave Theory and Techniques,* 2017, https://doi.org/10.1109/tmtt.2017.2677447.

[17] S. Ma, Y. Kumaresan, A. S. Dahiya, and R. Dahiya, "Ultra-Thin Chips with Printed Interconnects on Flexible Foils," *Advanced Electronic Materials,* vol. 8, no. 5, p. 2101029, 2022, https://doi.org/10.1002/aelm.202101029.

[18] R. A. Ramirez, T. M. Weller, M. Golmohamadi, J. Frolik, and J. Jamison, "Additive manufactured, on-package 2.4 GHz tripolar antenna system for cluttered channels," In *2021 IEEE 21st Annual Wireless and Microwave Technology Conference (WAMICON),* 28–29 April 2021, pp. 1–4, https://doi.org/10.1109/WAMICON47156.2021.9443598.

[19] S. L. Yu and E. A. Rojas-Nastrucci, "Characterization of microdispensed dielectric materials for direct digital manufacturing using coplanar waveguides," In *Presented at the 2019 IEEE 20th Annual Wireless and Microwave Technology Conference (WAMICON),* 2019.

[20] R. A. Ramirez, E. A. Rojas-Nastrucci, and T. M. Weller, "Laser-assisted additive manufacturing of mm-wave lumped passive elements," *IEEE Transactions on Microwave Theory and Techniques,* vol. 66, no. 12, pp. 5462–5471, 2018, https://doi.org/10.1109/TMTT.2018.2873294.

[21] E. A. Rojas-Nastrucci, R. A. Ramirez, and T. M. Weller, "Direct digital manufacturing of mm-wave vertical interconnects," In *2018 IEEE 19th Wireless and Microwave Technology Conference (WAMICON),* 9–10 April 2018, pp. 1–3, https://doi.org/10.1109/WAMICON.2018.8363917.

[22] M. M. Abdin, W. J. D. Johnson, J. Wang, and T. M. Weller, "W-band MMIC chip assembly using laser-enhanced direct print additive manufacturing," *IEEE Transactions on Microwave Theory and Techniques*, vol. 69, no. 12, pp. 5381–5392, 2021, https://doi.org/10.1109/TMTT.2021.3124237.

[23] N. J. Wilkinson, M. A. A. Smith, R. W. Kay, and R. A. Harris, "A review of aerosol jet printing-a non-traditional hybrid process for micro-manufacturing," *The International Journal of Advanced Manufacturing Technology*, vol. 105, no. 11, pp. 4599–4619, 2019, https://doi.org/10.1007/s00170-019-03438-2.

[24] F. Cai, Y. H. Chang, K. Wang, C. Zhang, B. Wang, and J. Papapolymerou, "Low-loss 3-D multilayer transmission lines and interconnects fabricated by additive manufacturing technologies," *IEEE Transactions on Microwave Theory and Techniques*, vol. 64, no. 10, pp. 3208–3216, 2016, https://doi.org/10.1109/tmtt.2016.2601907.

[25] C. Fan et al., "Aerosol jet printing for 3-D multilayer passive microwave circuitry," In *Microwave Conference (EuMC), 2014 44th European*, 6–9 Oct. 2014, pp. 512–515, https://doi.org/10.1109/EuMC.2014.6986483.

[26] C. Fan, C. Yung-Hang, W. Kan, W. T. Khan, S. Pavlidis, and J. Papapolymerou, "High resolution aerosol jet printing of D- band printed transmission lines on flexible LCP substrate," In *2014 IEEE MTT-S International Microwave Symposium (IMS)*, 1–6 June 2014, pp. 1–3, https://doi.org/10.1109/mwsym.2014.6848597.

[27] V. T. Bharambe, Y. Oh, J. J. Adams, D. Negro, and E. MacDonald, "3D printed zirconia for UWB stacked conical ring DRA," In *2020 IEEE International Symposium on Antennas and Propagation and North American Radio Science Meeting*, 5–10 July 2020, pp. 41–42, https://doi.org/10.1109/IEEECONF35879.2020.9330248.

[28] B. Mummareddy et al., "Mechanical properties of material jetted zirconia complex geometries with hot isostatic pressing," *Advances in Industrial and Manufacturing Engineering*, vol. 3, p. 100052, 2021, https://doi.org/10.1016/j.aime.2021.100052.

[29] Y. Oh et al., "Microwave dielectric properties of zirconia fabricated using nanoparticle jetting(™)," *Additive Manufacturing*, vol. 27, pp. 586–594, 2019, https://doi.org/10.1016/j.addma.2019.04.005.

[30] G. Addamo et al., "3-D printing of high-performance feed horns from Ku- to V-bands," *IEEE Antennas and Wireless Propagation Letters*, vol. 17, no. 11, pp. 2036–2040, 2018, https://doi.org/10.1109/LAWP.2018.2859828.

[31] B. Zhang and e. al., "Metallic 3-D printed antennas for millimeter- and submillimeter wave applications," *IEEE Transactions on Terahertz Science and Technology*, vol. 6, no. 4, pp. 592–600, 2016, https://doi.org/10.1109/tthz.2016.2562508.

[32] E. A. Rojas-Nastrucci, J. T. Nussbaum, N. B. Crane, and T. M. Weller, "Ka-band characterization of binder jetting for 3-D printing of metallic rectangular waveguide circuits and antennas," *IEEE Transactions on Microwave Theory and Techniques*, vol. 65, no. 9, pp. 3099–3108, 2017, https://doi.org/10.1109/TMTT.2017.2730839.

[33] E. A. Rojas-Nastrucci, J. Nussbaum, T. M. Weller, and N. B. Crane, "Meshed rectangular waveguide for high power, low loss and reduced weight applications," In *2016 IEEE MTT-S International Microwave Symposium (IMS)*, 22–27 May 2016, pp. 1–4, https://doi.org/10.1109/mwsym.2016.7540079.

[34] E. A. Rojas-Nastrucci, J. Nussbaum, T. M. Weller, and N. B. Crane, "Metallic 3D Printed Ka-Band Pyramidal Horn using Binder Jetting," In *Presented at the IEEE Latin America Microwave Conference 2016, Pto. Vallarta, MEX*, 2016.

[35] M. D'Auria et al., "3-D printed metal-pipe rectangular waveguides," *IEEE Transactions on Components, Hybrids, and Manufacturing Technology*, vol. 5, no. 9, pp. 1339–1349, 2015, https://doi.org/10.1109/tcpmt.2015.2462130.

[36] K. Zhao and D. Psychogiou, "Monolithic multiband coaxial resonator-based bandpass filter using stereolithography apparatus (SLA) manufacturing," *IEEE Transactions on Microwave Theory and Techniques*, vol. 70, no. 9, pp. 4156–4166, 2022, https://doi.org/10.1109/TMTT.2022.3193701.

[37] M. Husain, S. Jamshed, and N. Qureshi, "Transient aero-thermal analysis of high speed vehicles using CFD," In *Proceedings of 2012 9th International Bhurban Conference on Applied Sciences & Technology (IBCAST)*, 9–12 Jan. 2012, pp. 171–175, https://doi.org/10.1109/IBCAST.2012.6177549.

[38] C. Wang et al., "Coupling model and electronic compensation of antenna-radome system for hypersonic vehicle with effect of high-temperature ablation," *IEEE Transactions on Antennas and Propagation*, vol. 68, no. 3, pp. 2340–2355, 2020, https://doi.org/10.1109/TAP.2019.2948502.

[39] W. C. Wilson and G. M. Atkinson, "Passive wireless sensor applications for NASA's extreme aeronautical environments," *IEEE Sensors Journal*, vol. 14, no. 11, pp. 3745–3753, 2014, https://doi.org/10.1109/JSEN.2014.2322959.

[40] A. Parameswaran, H. S. Sonalikar, and D. Kundu, "Temperature-dependent electromagnetic design of inhomogeneous planar layer variable thickness radome for power transmission enhancement," *IEEE Antennas and Wireless Propagation Letters*, vol. 20, no. 8, pp. 1572–1576, 2021, https://doi.org/10.1109/LAWP.2021.3091516.

[41] J. Mei and Y. J. Xie, "Effects of a hypersonic plasma sheath on the performances of dipole antenna and horn antenna," *IEEE Transactions on Plasma Science*, vol. 45, no. 3, pp. 364–371, 2017, https://doi.org/10.1109/TPS.2017.2656159.

[42] A. Scarabosio et al., "Radiation and scattering of EM waves in large plasmas around objects in hypersonic flight," *IEEE Transactions on Antennas and Propagation*, vol. 70, no. 6, pp. 4738–4751, 2022, https://doi.org/10.1109/TAP.2022.3142310.

[43] R. Tang, Z. Xiong, K. Yuan, M. Mao, Y. Wang, and X. Deng, "EHF wave propagation in the plasma sheath enveloping sharp-coned hypersonic vehicle," *IEEE Antennas and Wireless Propagation Letters*, vol. 20, no. 6, pp. 978–982, 2021, https://doi.org/10.1109/LAWP.2021.3068575.

[44] B. Yao, L. Shi, X. Li, H. Wei, and Y. Liu, "Spatial selectivity of plasma sheath channel at millimeter wave band for hypersonic vehicle," *IEEE Transactions on Plasma Science*, vol. 50, no. 8, pp. 2430–2438, 2022, https://doi.org/10.1109/TPS.2022.3183470.

[45] M. C. Scardelletti, J. L. Jordan, and G. E. Ponchak, "Temperature dependency (25°C–400°C) of a planar folded slot antenna on alumina substrate," *IEEE Antennas and Wireless Propagation Letters*, vol. 7, pp. 489–492, 2008, https://doi.org/10.1109/LAWP.2008.2006068.

[46] H. Cheng, S. Ebadi, X. Ren, and X. Gong, "Wireless passive high-temperature sensor based on multifunctional reflective patch antenna up to 1050 degrees centigrade," *Sensors and Actuators A: Physical*, vol. 222, pp. 204–211, 2015, https://doi.org/10.1016/j.sna.2014.11.010.

[47] H. Cheng et al., "Evanescent-mode-resonator-based and antenna-integrated wireless passive pressure sensors for harsh-environment applications," *Sensors and Actuators A: Physical*, vol. 220, pp. 22–33, 2014, https://doi.org/10.1016/j.sna.2014.09.010.

[48] S. L. Yu et al., "Advanced manufacturing of passive wireless high-temperature pressure sensor using 3-D laser machining," In *2022 IEEE 22nd Annual Wireless and Microwave Technology Conference (WAMICON)*, 27–28 April 2022, pp. 1–4, https://doi.org/10.1109/WAMICON53991.2022.9786160.

[49] S. Wang et al., "3-D printed zirconia ceramic Archimedean spiral antenna: theory and performance in comparison with its metal counterpart," *IEEE Antennas and Wireless Propagation Letters*, vol. 21, no. 6, pp. 1173–1177, 2022, https://doi.org/10.1109/LAWP.2022.3161004.

[50] S. Wang et al., "Radar cross-section reduction of helical antenna by replacing metal with 3-D printed zirconia ceramic," *IEEE Antennas and Wireless Propagation Letters*, vol. 19, no. 2, pp. 350–354, 2020, https://doi.org/10.1109/LAWP.2019.2962524.

[51] Y. Oh, V. T. Bharambe, J. J. Adams, D. Negro, and E. MacDonald, "Design of a 3D printed gradient index lens using high permittivity ceramic," In *2020 IEEE International Symposium on Antennas and Propagation and North American Radio Science Meeting*, 5–10 July 2020, pp. 1431–1432, https://doi.org/10.1109/IEEECONF35879.2020.9330193.

[52] Y. Oh, N. Kordsmeier, H. Askari, and J. J. Adams, "Low profile GRIN lenses with integrated matching using 3-D printed ceramic," *IEEE Open Journal of Antennas and Propagation*, vol. 4, pp. 12–22, 2023, https://doi.org/10.1109/OJAP.2022.3227747.

[53] C. Neff, E. Elston, and A. Schrand, "Interconnections for additively manufactured hybridized printed electronics in harsh environments," *Designs*, vol. 4, no. 2, pp. 14, https://doi.org/10.3390/designs4020014.

[54] S. L. C. Yu, Christopher Polotai, Anton and E. A. Rojas-Nastrucci, "Laser enhanced direct-print additive manufacturing (LE-DPAM) of mm-wave antenna using LTCC dielectric paste for high temperature applications," In *Presented at the 2023 IEEE 23rd Annual Wireless and Microwave Technology Conference (WAMICON)*, 2023.

[55] N. Dib, T. Weller, M. Scardelletti, and M. Imparato, "Analysis of cylindrical transmission lines with the finite-difference time-domain method," *IEEE Transactions on Microwave Theory and Techniques,* vol. 47, no. 4, pp. 509–512, 1999, https://doi.org/10.1109/22.754886.

[56] W. Kuang-Yuh and J. Kauffman, "Radiation pattern computations for cylindrical-rectangular microstrip antennas," In *1983 Antennas and Propagation Society International Symposium*, 23–26 May 1983, vol. 21, pp. 39–42, https://doi.org/10.1109/APS.1983.1149119.

[57] N. Herscovici, Z. Sipus, and P. S. Kildal, "The cylindrical omnidirectional patch antenna," *IEEE Transactions on Antennas and Propagation*, vol. 49, no. 12, pp. 1746–1753, 2001, https://doi.org/10.1109/8.982455.

[58] X. Zhang, X. Ma, and Q. Lai, "Two kind of conical conformal GPS antenna arrays on projectile," In *2009 3rd IEEE International Symposium on Microwave, Antenna, Propagation and EMC Technologies for Wireless Communications*, 27–29 Oct. 2009, pp. 659–662, https://doi.org/10.1109/MAPE.2009.5355532.

[59] A. I. Zaghloul, T. K. Anthony, and W. O. Coburn, "A study on conical spiral antennas for UHF SATCOM terminals," In *Proceedings of the 2012 IEEE International Symposium on Antennas and Propagation*, 8–14 July 2012, pp. 1–2, https://doi.org/10.1109/APS.2012.6349035.

[60] V. Jaeck et al., "A conical patch antenna array for agile point-to-point communications in the 5.2-GHz band," *IEEE Antennas and Wireless Propagation Letters*, vol. 15, pp. 1230–1233, 2016, https://doi.org/10.1109/LAWP.2015.2502724.

[61] V. Jaeck et al., "A switched-beam conformal array with a 3-D beam forming capability in C-Band," *IEEE Transactions on Antennas and Propagation*, vol. 65, no. 6, pp. 2950–2957, 2017, https://doi.org/10.1109/TAP.2017.2696418.

[62] A. F. Tinoco-S, P. C. Ribeiro-Filho, M. V. T. Heckler, J. C. D. S. Lacava, and O. M. C. Pereira-Filho, "Fast predesigning of circumferential arrays of probe-fed microstrip antennas," In *2016 10th European Conference on Antennas and Propagation (EuCAP)*, 10–15 April 2016, pp. 1–4, https://doi.org/10.1109/EuCAP.2016.7481525.

[63] S. Xiao, S. Yang, H. Zhang, Q. Xiao, Y. Chen, and S. W. Qu, "Practical implementation of wideband and wide-scanning cylindrically conformal phased array," *IEEE Transactions on Antennas and Propagation*, vol. 67, no. 8, pp. 5729–5733, 2019, https://doi.org/10.1109/TAP.2019.2922760.

[64] J. Santiago-Prowald and L. S. Drioli, "Space environment and materials," In *Space Antenna Handbook*, John Wiley & Sons, Ltd. 2012, pp. 106–132.

[65] D. M. Hurley et al., "An analytic function of lunar surface temperature for exospheric modeling," *Icarus*, vol. 255, pp. 159–163, 2015, https://doi.org/10.1016/j.icarus.2014.08.043.

[66] J. I. Sahr et al., "Wireless Payload Thermal-Vacuum Testing for Lunar Harsh Environment," In *2021 IEEE Space Hardware and Radio Conference (SHaRC)*, 17–22 Jan. 2021, pp. 13–16, https://doi.org/10.1109/SHaRC51853.2021.9375873.

[67] R. Romanofsky, M. Scardelletti, and J. Jordan, "physical layer communications technologies: harsh planetary environments," *IEEE Microwave Magazine*, vol. 22, no. 12, pp. 73–85, 2021, https://doi.org/10.1109/MMM.2021.3110157.

[68] *General Environmental Verification Standard (GEVS) for GSFC Flight Programs and Projects*, GSFC-STD-7000, NASA, 2021.

[69] W. A. Imbriale, S. Gao, and L. Boccia, *Space Antenna Handbook*. New York, United Kingdom: John Wiley & Sons, Incorporated, 2012.

[70] R. E. Hodges, N. Chahat, D. J. Hoppe, and J. D. Vacchione, "A deployable high-gain antenna bound for mars: developing a new folded-panel reflectarray for the first cubesat mission to mars," *IEEE Antennas and Propagation Magazine*, vol. 59, no. 2, pp. 39–49, 2017, https://doi.org/10.1109/MAP.2017.2655561.

[71] N. Chahat, *CubeSat antenna design*, First edition. ed. Hoboken, New Jersey: Wiley-IEEE, 2020.

[72] N. Miguélez-Gómez et al., "Thickness-accommodation in X-band origami-based reflectarray antenna for small satellites applications," In *2020 IEEE International Conference on Wireless for Space and Extreme Environments (WiSEE)*, 12–14 Oct. 2020, pp. 54–59, https://doi.org/10.1109/WiSEE44079.2020.9262670.

[73] O. A. Peverini et al., "Integration of an H-plane bend, a twist, and a filter in Ku/K-band through additive manufacturing," *IEEE Transactions on Microwave Theory and Techniques*, vol. 66, no. 5, pp. 2210–2219, 2018, https://doi.org/10.1109/TMTT.2018.2809505.

[74] E. A. Rojas, J. T. Nussbaum, T. M. Weller, and N. B. Crane, "Apertured waveguides for electromagnetic wave transmission," ed: Google Patents, 2019.

[75] S. Sirci, E. Menargues, and M. Billod, "Space-qualified additive manufacturing and its application to active antenna harmonic filters," In *2021 IEEE MTT-S International Microwave Filter Workshop (IMFW)*, 17–19 Nov. 2021, pp. 239–242, https://doi.org/10.1109/IMFW49589.2021.9642306.

[76] N. Chamberlain, M. S. Barbetty, G. Sadowy, E. Long, and K. Vanhille, "A dual-polarized W-band metal patch antenna element for phased array applications," In *2014 IEEE Antennas and Propagation Society International Symposium (APSURSI)*, 6–11 July 2014, pp. 1640–1641, https://doi.org/10.1109/APS.2014.6905146.

[77] H. Kazemi et al., "Ultra-compact G-band 16way power splitter/combiner module fabricated through a new method of 3D-copper additive manufacturing," In *2015 IEEE MTT-S International Microwave Symposium*, Phoenix, AZ, 17–22 May 2015, pp. 1–3, https://doi.org/10.1109/mwsym.2015.7166718.

4 Printed Pressure Sensors for Extreme Environments

Tosin D. Ajayi
North Carolina State University

Spencer Nguyen
Florida State University

Amanda Schrand
Air Force Research Laboratory

Chengying Xu
North Carolina State University

4.1 INTRODUCTION

The growing use of advanced materials such as ceramics, polymers, and composites for high-performance applications ranging from aerospace to energy, environmental, and defense-oriented applications push on for the deployment of rapid and reliable manufacturing techniques. For example, advanced ceramics such as ultra-high-temperature ceramics (UHTCs) are usually considered high-performance structural materials because of their superior high temperature/chemical stability and strength [1–3]. However, the fabrication with these high-performance materials to suit a system is usually difficult due to various assembly needs, which is significantly challenging for functional materials used in sensors. For instance, one major need is the use of defined surfaces in hard tooling, such as mold cavities, to aid the shaping of ceramics geometries to obtain net-shaped three-dimensional (3D) components, such as during dry pressing and tape casting processes [4,5]. Another need is post-processing such as post-hard machining, also known as subtractive manufacturing, of the individually fabricated-material parts, before assembly of those parts to form a system. These requirements make the fabrication and assembly processes of these systems time-consuming and expensive [6]. Thus, the difficulty in manufacturing complex-shaped components without use of molds/dies, hard tooling, or any other traditional means has brought about the creation and growth of novel manufacturing techniques such as additive manufacturing (AM) which is also known as 3D printing.

DOI: 10.1201/9781003138945-4

4.1.1 BRIEF INTRODUCTION TO ADDITIVE MANUFACTURING (AM)

According to ASTM F2792 Standard Terminology for Additive Manufacturing Technologies [7], AM is defined as a process of joining materials to make objects from 3D model data, usually layer-upon-layer, as opposed to subtractive manufacturing methods. The manufacturing process is sub-divided into two: (1) a computer-aided design (CAD) model is used to process the geometrical information of the parts through a slicing algorithm (STL file) that draws detailed information for every layer, (2) a computer-controlled machine/equipment where the feedstock (start-up materials) is typically fed in the form of powder, suspension, or paste and deposited layer by layer, point by point, or line by line to make desired parts. Each layer is usually of even thickness that is subsequently stacked and fused together through binding, curing, and/or sintering. The physical size of the fabricated parts may range from microscopic such as printed electronic components to macroscopic such as an entire building [8]. In general, characteristics of AM technology contain the following [9]:

- The geometry of each part is generated only from the CAD model.
- The material properties of the part are obtained during the buildup process.
- There is no need of tooling process such as machining or use of molds and dies for the shaping process.
- Parts can be built without the need for clamping, thereby eliminating the clamping challenge associated with subtractive manufacturing techniques.

Initially, AM was created as a method for rapid prototyping (RP), where individual products are manufactured without any requirement for a dedicated tooling. The RP technology allowed preparation of parts at a faster rate at a single run and thus reducing the time span between designs, testing, and implementation. Then, during the 1980s, the basic technologies for AM were developed for the purpose of adding design flexibility to manufacturing process. However, overtime, the terminology for this class of technologies shifted gradually from RP to AM [10]. More recently, the terms "additive manufacturing" and "3D printing" are often used synonymously, which are basically defined as the joining of materials under computer control to create 3D objects from digital data.

4.1.2 PRESSURE SENSORS

Pressure monitoring is of significant importance especially in environments where real-time information to understand and control processes is required [11,12], such as gas pipelines pressure metering [13], turbines [14], and downhole pressure monitoring [15]. Usually, pressure monitoring is under harsh environments, for example, high-temperature conditions and time-varying aqueous environments. Consequently, it is imperative for pressure sensors to be capable of operating at high temperatures and harsh environments (e.g., >500°C). For instance, in an aircraft engine, air-pressure measurement as the air approaches the engine's combustor helps to control the air/fuel mixture to reduce the risk of stall. This requires a sensor that operates in a turbulent gas environment at elevated temperatures from 600°C to 1,800°C [16].

Most commercially available pressure sensors can be categorized into three main groups [17]: (1) membrane-based, (2) piezoelectric effect, and (3) solid-state sensors. A typical membrane-based pressure sensor is composed of a flexible silicon membrane as the sensing element and incorporated with silicon piezoresistors or capacitors for data retrieval [18]. However, these devices are typically not suitable for extremely high-temperature environments due to materials and readout limitations. Piezoelectric pressure sensors leverage the piezoelectric effect to measure changes in pressure, acceleration, temperature, strain, or force by converting them to an electrical charge [19,20]. Previous studies show that piezoelectric pressure sensors yield reliable results because piezoelectric devices have a high-frequency response and can achieve excellent signal conversion without requiring any bellows, diaphragm, or any type of mechanical linkage. However, piezoelectric sensors are suitable only for measuring dynamic pressure changes due to the inherent signal drift of these devices [17]. Finally, solid-state pressure sensors are fabricated typically on a rigid surface through complementary metal oxide semiconductor (CMOS) processes [21–23] and make use of hanging structures such as diaphragm and cavity-based systems, thereby rendering them expensive to fabricate. Moreover, the rigidity of these devices makes them unsuitable for applications where flexibility is desired.

Considering the aforementioned, it becomes imperative to look for alternative fabrication processes that can deliver cost-effective, flexible, and conformal pressure sensors to overcome the limitations associated with conventional fabrication processes. The quest to solve these challenges has led researchers and scientist exploring printing technology.

The fabrication of sensing devices through traditional printing techniques has been gaining increasing attention in recent years. Some of the advantages associated with printing methods include improved cost efficiency, reduction of wastage of material during fabrication, flexibility in the substrate, and low manufacturing temperatures [24]. Based on their respective fabrication techniques, the 3D printing process was originally grouped into seven main categories according to standard terminology for AM technologies (ASTM F2792-12a): vat polymerization, material jetting, binder jetting, material extrusion, powder bed fusion, sheet lamination, and directed energy deposition. In this chapter, we focus mainly on the techniques of 3D printing that have been used to fabricate some pressure sensors, as well as discuss the advances and limitations in different fabricating processes, and then describe a number of investigated devices.

4.2 PRINTING TECHNIQUES FOR PRESSURE SENSORS FABRICATION

4.2.1 FUSED DEPOSITION MODELING

Fused Deposition Modeling (FDM) is an example of common material extrusion process and a bulk-solid based AM technique which was developed and commercialized by Stratasys Inc. in Eden Prairie, Minnesota [25]. This solid freeform fabrication method is done by building a 3D object from a filament, which is fed into an extruder head capable of moving in the X–Y–Z directions, where the filament material is selectively extruded from nozzles. This method is generally used to fabricate polymeric components, but the technology has since been extended to the fabrication of ceramic components. In this case, a dried surfactant pre-coated ceramic powder mixed with appropriate volume of binder agent is thoroughly mixed with thermoplastic polymers. The product is then extruded to form filaments via a pair of counter-rotating rollers [26,27].

The filament is fed into the heated extrusion (resistive heater) system and then extruded through a nozzle. The molten feedstock is then deposited in the X_Y plane on a Z stage platform to obtain the ceramic green part. After each layer is completed, the part is lowered by one-layer thickness and the process repeats. Afterwards, the green product is exposed to high temperature for sintering to remove the binder and then densify the component. The method of modifying the FDM method to produce ceramic components is sometimes termed as fused deposition of ceramics (FDC). Generally, the FDC process can be categorized into four stages [28]:

   i. Optimization of thermoplastic binder composition.
   ii. Fabrication of the filament.
   iii. Process fabrication (systematic printing of layers).
   iv. Binder removal and sintering of the green body.

4.2.1.1 Filament Selection for Fused Deposition Modeling

In selecting the filament material feedstock, careful consideration has to be given to some important requirements. For instance, to prepare ceramic components, fine particles of ceramic with wide size distribution are required to improve the filament by decreasing the overall viscosity of the binder and powder. Likewise, the binder system should possess properties such as high mechanical strength, low viscosity, and high strain. For example, low binder viscosity helps in preventing the return of the molten material during the process. Another important requirement is the selection of adequate nozzle size and type for the extrusion of the filament in order to ensure proper flow. The layer thickness and vertical dimensional accuracy is determined by the extruder die diameter which ranges from 0.013 to 0.005 inches (0.3302–0.1270 mm) [29]. The required pressure drops ΔP to extrude a filament in FDC process are dependent on the viscosity of the feedstock, geometry of nozzle, nozzle orifice diameter, and the volumetric flow rate. Here, pressure drop can be assumed to be the relevant pressure on filament, when described in absolute pressure terms. Therefore, if this pressure exceeds the critical load per unit area for the filament, buckling happens [30]. The buckling criterion is given by Euler's analysis for on-ended boundary condition:

$$\sigma_{cr} = \frac{\pi^2 E}{4(L/R)^2}$$

where σ_{cr} is the critical buckling stress, E is the compressive modulus for FDC filaments, L is the length of the filament between the roller and the top of the liquefier, and R is the radius of the filaments. Also, the pressure drop ΔP can be estimated by assuming a non-Newtonian fluid moving in a capillary rheometer having a length l and radius r which is given as follows:

$$\Delta P = \frac{8\eta_a Q l}{\Pi r^4}$$

where η_a is the apparent viscosity obtained via a capillary rheometer and Q is the volumetric flow rate. Therefore, filaments in FDC will buckle when extrusion pressure exceeds critical buckling stress, i.e.,

$$\Delta P' > \sigma_{cr} \text{ or } \frac{\Delta P}{k} > \sigma_{cr}$$

where $\Delta P'$ and ΔP are the FDC extrusion pressure and capillary extrusion pressure, respectively. k is the scaling factor between FDC extrusion pressure and capillary extrusion pressure, respectively, for a particular filament material. Therefore, from the above equations, the FDC filament will buckle if:

$$E / \eta_a < \frac{8 Q l (L/R)^2}{\pi^3 r^4 k}$$

In general, feedstocks for filaments extrusion should have the following requirements [31]:

i. Homogenous particle distribution to ensure constant flow and avoid flaws due to agglomerates.
ii. Ability to counteract shrinkage and cracking during sintering or binder burnout.
iii. Solvent migration and sedimentation of particles should be avoided to ensure constant flow and storage stability.
iv. Rheology during the process should support good shape retention and good adhesion of the filaments.
v. Suitable particle size and distribution for the applied nozzle dimensions.
vi. The binder phase should allow ease of drying and burning out of binder.
vii. Suitable solidification kinetics such as vapor pressure, thixotropy, thermal conductivity, heat capacity, and melting points.
viii. Suitable interface properties for adhesion and fusion of the filaments.

4.2.1.2 Fused Deposition Modeling Printed Pressure sensors

Shemelya et al. [32,33] designed and developed capacitive series of sensors that are fully encapsulating a copper wire, a copper mesh, or a combination of both including some other electronic devices. In one design, a fully encapsulated capacitive sensor was fabricated using polycarbonate (PC) printed from a Stratasys FDM Titan printer. The conductive materials used as sensors in this device are copper wire (320 μm dia.) and copper mesh (149 μm mesh spacing, 228 μm wire dia.) and were laser welded to a microcontroller. Other electronic devices including a Zener diode, LEDs, resistors,

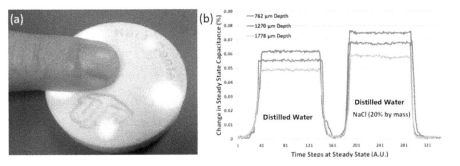

FIGURE 4.1 (a) A miniature capacitive sensor encapsulated in polycarbonate. (b) Detection of distilled H_2O and $NaCl:H_2O$ (20% by mass) with a FDM copper mesh capacitance sensor [32,33].

and electrical connections for this sensor were fully encapsulated in the PC by the FDM Titan printer, thereby further demonstrating the feasibility of integrating fully embedded bulk conductors into a FDM process. To display the capacitive sensing functionality, the sensor was used in a microfluidic mixing device. The microfluid mixing device was fabricated with a polyphenylsulfone (PPSF) also from a Stratasys Titan printer. The sensor was used to identify two visually identical liquids, distilled water and a 20% by mass $NaCl:H_2O$ solution (Figure 4.1). The distinction was made effectively at sensor depths greater that 1,500 µm. The results indicate that the sensors have potential applications in liquid flow and microfluidic sensing.

Saari et al. [34] used a combination of fiber encapsulation additive manufacturing (FEAM) and thermoplastic elastomer additive manufacturing (TEAM) which are types of FDM to develop a capacitive force sensor. This capacitor is made up of two electrically conductive outer plates, a dielectric, and embedded wires. The capacitor's plates were printed using acrylonitrile butadiene styrene (ABS) and the dielectric from a thermoplastic elastomer using the TEAM process from a FDM filament extruder called X-Truder. The electrodes were printed using the FEAM process with a copper wire to form square planar coils having an outside dimension of 24 by 24 mm and pitch of 0.38 mm. The wires were connected in a tight spiral form and the dielectric is compressed under load between the two plates. The sensing mechanism is because of the distance between the plates due to an applied load. The load testing result of the device displayed good synchronization between the load and capacitance data except for 8.3 second delay in the capacitance measurement during unloading hypothesized to be caused by the material hysteresis (Figure 4.2).

4.2.2 Ultrasonic Additive Manufacturing (UAM)

Hehr et al. from Fabrisonic LLC [35] designed and fabricated a new fiber Bragg grating (FBG) based strain sensor by utilizing the FBG with metallic structure. They used a method known as ultrasonic additive manufacturing (UAM), which allows for a low-temperature 3D printing of metals to overcome the challenge of high formation temperature during regular 3D metal printing. High formation

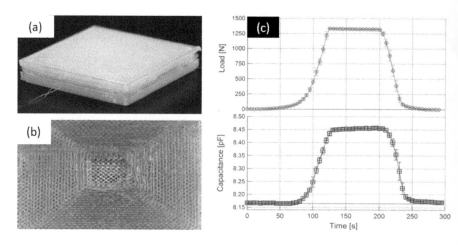

FIGURE 4.2 (a) Image of complete printed sensor. (b) Photo of internal spiral structure of sensor. (c) Average load and capacitance vs. time for sensor [34].

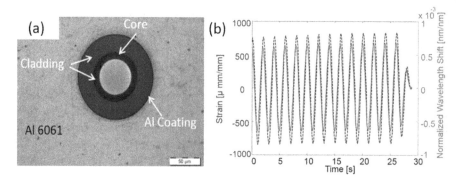

FIGURE 4.3 (a) Metallized fiber optic cables. (b) Exterior strain and embedded FBG wavelength shift (not calibrated strain) cyclic profile comparison [35].

temperature causes FBGs to crack due to high thermal stress. The UAM process involves building up a solid metal part by ultrasonically welding a succession of metal tapes. The ultrasonic welding occurs through scrubbing of metals together which generates enough localized heat at the interface for joining – approximately 150°C for aluminum. To avoid sensor damage, a precut was made on the aluminum. The fiber (Ormocer®, 195 μm diameter, grating length of 10 mm) was then placed into the precut channel and welded over with five layers of AA 6061 H18 foil with a Fabrisonic SonicLayer™ 7,200 printer. The foil thickness used to consolidate the fiber is 0.15 mm. When tested in the tensile direction, the specimen exhibited uniform axial strain field which is equal inside and outside of the specimen. The embedded sensors did not show sign of slip behavior until after the yield point of the material (Figure 4.3).

4.2.3 SELECTIVE LASER PROCESSES

Selective laser sintering (SLS) and selective laser melting (SLM) are a type of powder bed fusion process that involves the selective use of a high-power laser beam to irradiate the surface of target powder bed [36]. This method is commonly used to print 3D ceramic and metallic products. The powder is heated up and sintered through the fusion between the particles. Subsequently, a new layer of powder is spread on the previous surface, and the heating process is repeated. Thinner layers with powders of small particles sizes are recommended to achieve finer details. Although finer particles mean the layer is susceptible to higher reactivity with oxygen and humidity, an appropriate approach should be taken to prevent oxidation. Additionally, powder agglomeration should be prevented through efficient approaches. In summary, the heating process in SLS and SLM can be summarized as follows:

    i. Input of energy and absorption
    ii. Powder bed heating
    iii. Melting and sintering

4.2.3.1 Selective Laser Processing Parameters

Selective laser systems usually consist of three major parts: laser system, powder bed, and spreading system [37]. The laser system is made up of the laser and scanner. The laser beam is focused on the target powder surface and then absorbed by the powder for melting to take place, whereas the scanner allows the laser to move in a 2D plane. After laser scan is over, the temperature (T_m) drops, viscosity is increased, and diffusion rate decreases as the part is formed layer by layer. The fill power (P) can be described with the following equation [38]:

$$P = \frac{BS * \rho * D_b * h * \left[C * (T_m - T_b) + l_f \right]}{(1 - R)}$$

where BS is the beam speed, ρ is the power density, D_b is the diameter of beam on laser bed, h is the slice thickness, C is the specific heat, T_m is the melting temperature, T_b is the bed temperature, l_f is the latent melting temperature, and R is the reflectivity (fraction of laser light reflected by powder).

Powders also play an important role in having a successful SLS process. For instance, powder size influences laser melting efficiency. Large particles usually need more energy to melt and also reduce the pack density of the powder bed [37]. But smaller particles are prone to agglomeration which can result in poor powder coating or layering. Therefore, particle size is recommended in the range of few to several hundred microns. Another important powder parameter is flowability, which is influenced by particle morphology and size distribution. For instance, powders with spherical orientation and narrow size distribution flow better [39]. Therefore, powders of good flowability are therefore required to have a uniform thickness and flat surface, which in turn allows for uniform laser absorption. The powder layer is also an important parameter that determines the laser beam penetration to the powder. The powder layer also influences the energy density to be applied and subsequently the surface roughness. The energy density is determined by [38]:

$$E = \frac{P * f}{BS * SP}$$

where E is the energy density (cal/cm^2), P is the power density, f is a conversion factor, and SP is the scan spacing.

4.2.3.2 Selective Laser Printed Pressure Sensors

Zhang et al. [40] fabricated a flexible pressure sensor via SLS process. The sensor was made of polydimethylsiloxane (PDMS) film coated with reduced oxide graphene (rGO) and a flexible interdigital electrode (IDE). The PDMS was prepared from a mold printed by SLS technology in which the nylon powders were sintered layer by layer under laser illumination to form the desired product. Afterwards, the PDMS mixture was poured into the 3D printed mold for casting. The rGO was deposited on the surface of the casted PDMS via spray coating technique. The flexible IDE was obtained from silver nanowires drop-coated on the PDMS substrate and patterned using laser etching process. The fabricated sensor showed excellent sensitivity of 55 kPa^{-1} and linearity (R^2 > 0.997) in a pressure range up to 100 kPa and decent sensitivity over 10 kPa^{-1} in range of 100–400 kPa. These results were attributed to its surface irregularity and roughness from the sintering process (Figure 4.4).

A soft somatosensory detecting sensor made from mixture of nylon and graphite powders was fabricated via SLS process by Wei et al. [41]. The mixed powder was

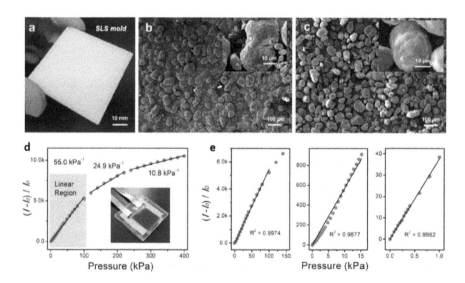

FIGURE 4.4 (a) Photo of 3D-printed mold from SLS process. SEM images of (b) surface of SLS-printed mold and (c) powders used for SLS printing. (d) Variation of the relative current $(I-I_o)/I_o$ vs. pressure and different sensitivity from different pressure ranges for the SLS sensor. (e) Performance of the SLS sensor in response to different pressure ranges [40].

spread in a powder bed and sintered with a laser (Coherent C series 30 W laser). The laser beam is directed on the surface of the powder layer through a galvanometer (effective aperture of 10 mm) and F–θ lens (focus length = 160 mm). At a temperature above its melting point, the polymer melts, flows, and soaks the graphite particles to form a soft hybrid conductive material. Since the graphite particles have a higher melting point compared to the polymer, its structure remained virtually almost the same. The fabricated pressure sensor displayed a stable response with good repeatability and sensitivity to movement from the human body (Figure 4.5).

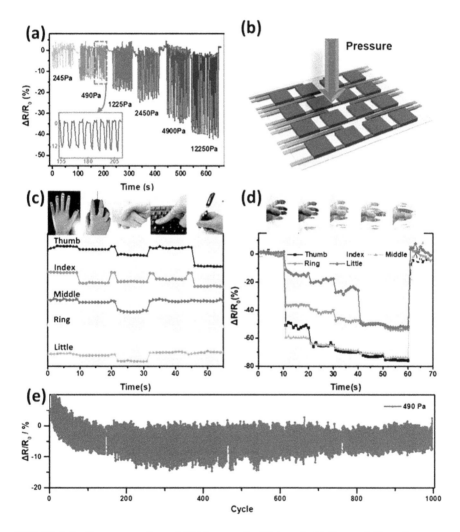

FIGURE 4.5 Pressure testing of SDM sensor by Wei et al. (a) Repeatability of pressure sensing. (b) Schematic of a 4-by-4 sensor array. (c and d) Corresponding electrical signal of the sensor-based pressure sensing gloves and in a 500 ml beaker of water with different volume. (e) Durability test of pressure sensor tested under a 490 Pa cyclic pressure with a frequency of 0.25 Hz [41].

Petrat et al. [42] integrated light emitting diode (LED) into a component of 316L chrome-nickel steel using combination of SLM and laser metal deposition (LMD). First, a base plate was manufactured with SLM which is made up of a solid body and an area with grid structure. The solid body includes a duct in the shape of a groove with a channel or path for carefully laying the power cable. The LED is embedded in the grid structure via an inlet from the solid body. The laser power used was 250 W with a scan speed between 700 and 1,400 mm/min for the SLM process. To completely embed the LED, the part was further filled with LMD using a TRUMPF TruDisk 2.0 kW Yb:YAG laser and a coaxial annular gap nozzle. The grain size of the 316L filler material ranges from 45 to 90 µm. Two filling strategies were used and compared with one another with respect to number of tracks and duration of welding time. The results were promising with regard to embedding electronic components into a metal part (Figure 4.6).

4.2.4 Inkjet Printing (IJP)

Inkjet printing (IJP) is a common AM technique that uses inkjet head for processing materials. This method can be used to fabricate components by ejecting liquid-phase materials (ink) in droplets form from the printhead nozzle onto substrates [43]. IJP process can be used in two modes: continuous or drop-on-demand (DOD), with the latter being the most used. The DOD mode gives opportunity to print components with higher positioning accuracy and smaller droplet size, which can occur via either thermal excitation or piezoelectric effect [44]. The major difference between the two

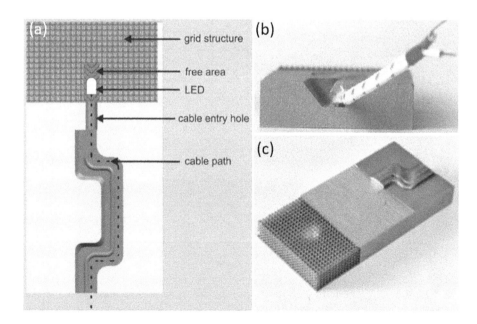

FIGURE 4.6   (a) Implementation overview of the LED in SLM part. (b) Positioned cable inside the duct before LMD process. (c) Component with successfully embedded LED [42].

is performance. In thermal systems, there is the presence of a heating element as a thin-film resistor such that the application of electric pulse at the head causes current to pass through the resistor, and the fluid in contact is vaporized and forms a bubble over the resistor. This bubble continues to expand in the fluid reservoir, and with continuous pressure, then a droplet is forced out of the nozzle unto the substrate [45]. In the piezoelectric system, as voltage is applied to the piezoelectric material coupled to the fluid, the fluid reservoir experiences a volumetric change. This change leads to a pressure buildup in the fluid that then causes a droplet to be ejected from the nozzle [46].

4.2.4.1 Important Processing Parameters in IJP

Varieties of materials such as ceramics and polymers can be processed using the IJP technique if the suspension (ink) is adequately prepared. A quality ink should display good printing performance and have excellent compatibility with the printer. The ink's printability can be estimated by [36,47,48]:

$$Z = \frac{1}{Oh} = \frac{Re}{\sqrt{We}} = \frac{(\gamma\rho a)^{1/2}}{\eta}$$

where Oh, Re, and We are the Ohnesorge number, the Reynolds number, and Weber number, respectively.

$$Re = \frac{\upsilon\rho a}{\eta}$$

$$We = \frac{\upsilon^2\rho a}{\gamma}$$

where a is the characteristic length representing the radius of the nozzle; υ, γ, ρ, and η are the ejection velocity, surface tension, density, and viscosity of the ink, respectively. The composition (powders, binders, etc.) of the suspension also plays an important role in having a successful printing operation. The materials used as binders must have suitable properties to prevent spreading from nozzles. The flowability and wettability of the particles are also important to prepare a printable suspension.

4.2.4.2 Inkjet-Printed Pressure Sensor

Lo et al. printed a soft, wearable pressure sensor patch via the IJP process [49]. The device is made of silver nanoparticle layer printed directly onto a PDMS substrate and then encapsulated by a very high bond (VHB) tape. First, the silver nanoparticles (100–200 nm particle size) were diluted in ethylene glycol to form the suspension. This suspension was then printed on a PDMS substrate using a GIX Microplotter (Sonoplot, Inc.) with nozzle openings of 50–200 μm. To aid the wetting of the ink, the PDMS substrate was treated with oxygen plasma at 30 W for 5 seconds. The authors demonstrated that the printed soft sensor patch, due to its high sensitivity, can measure arterial pulse waveforms or detect acoustic vibrations under various sound pressure levels (Figure 4.7).

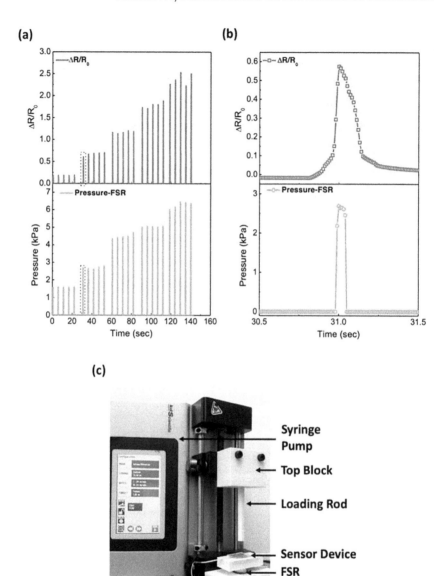

FIGURE 4.7 Dynamic pressure test of the sensor under varying loads. (a and b) Plots of relative changes in resistance for the pressure sensor from 1.5 to 6.5 kPa at a frequency of 0.2 Hz. The data in (b) is a zoomed in view of the data of dashed box in (a). (c) Photo of the experimental setup for dynamic pressure test on sensor [49].

4.3 OTHER AM METHODOLOGIES FOR FABRICATING ADVANCED PRESSURE SENSORS

Apart from the three main methods (fused deposition method, selective laser process, and IJP) described above, other AM technologies have also been used to fabricate pressure-based sensors. In one work, Santiago et al. [50] demonstrated the use of vat polymerization technique to fabricate a sensorized elastomer for wearable applications. Their method involved carefully weaving a pair of wires in adjacent unit-cell layers through a lattice of a proprietary polymer material in order to serve as a complex capacitor. The dimension of the single hexagonal lattice used is 50 mm×50 mm×24 mm which also acts as a self-aligning collection point for the wires. This work was aimed at improving the performance of American football helmets. When the structure is under load, the capacitance increases as the lattice is compressed, and lattice deformation is detected. Results from both static compression as well as low-velocity impact confirm the usability of the structure for wearable applications (Figure 4.8).

In another work, Juhasz et al. [51] used hybrid AM by combining directed energy deposition (DED) and traditional subtractive processing (such as machining) to fabricate high-performance metallic structures with embedded printed strain sensors. Hybrid manufacturing allows the development of advanced intricate structural elements within a single build envelope. The sensors were printed from functional ceramic-based, thick-film inks. To ensure that the sensors withstand the laser cladding temperature during DED, a steel plate was printed with a cavity to house the sensor before placing in the metallic structure. The sensor holder was created with the same laser cladding process and materials as the metallic structure or steel bar. After the sensor holder has been placed in the steel bar, a laser cladding consolidation process followed which involved stitch welding, gap deposition, cleanup machining, final deposition, and finish machining. The laser cladding system used in this work is called an Ambit™ tool from Youngstown State University, which specifically was

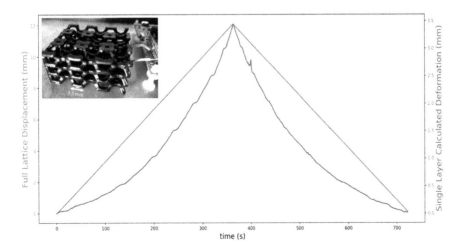

FIGURE 4.8 Quasi-static compression (2 mm/min) to 500 N. Inset is the image of the elastomer lattice with electronics [50].

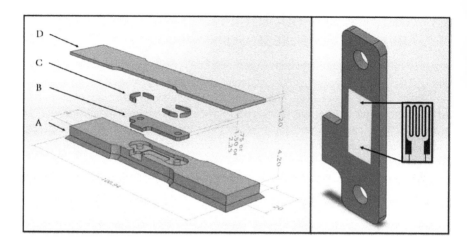

FIGURE 4.9   (Left side) Schematic of process for integrated strain sensor: (A – dark green) main laser clad section of tensile bar; (B – blue) laser clad and machined sensor plate with screen printed piezoresistive strain gauge on bottom surface; (C – brown) gap filling laser cladding for removing voids; (D – light green) final laser cladding and machining to consolidate the entire structure and encapsulate the sensor plate. (Right side) Enlarged view of sensor plate with schematic of screen printed sensor and location [51].

developed to provide ease of integration of additional tooling with an exchange carousel that can provide tools for robotic placement of components, micro-dispensing of functional thick-film inks, or polymer deposition. A tensile bar built from this process was used to evaluate the performance of the deposition by collecting data from the internal sensor, and one result showed that the embedded sensors successfully responded to cyclic applied stress (Figure 4.9).

4.4   PRINTED SENSORS THROUGH POLYMER-DERIVED CERAMIC (PDC) PROCESS

The existing methods for remote sensing (i.e., fuel sensor, accelerometer, surface acoustic wave sensor, chemical resistor, temperature sensor, etc.) are performed by proximate environmental monitoring and are primarily limited to storage and transportation purposes [52]. However, the development of next-generation defense systems demands a greater situational awareness of the extreme environmental conditions such as high-temperature and high-pressure environments. Therefore, novel sensors that can increase the safety and effectiveness of critical materials by tracking and assessing their status on demand, and over extended periods of time, based upon extreme and transient conditions will find great utility in many defense applications.

4.4.1   Printed Polymer-Derived Ceramics

Polymer-derived ceramics (PDC) have been gaining a lot of attention for the past 40 years as an exceptionally formidable tool to manufacture advanced ceramics [53]. The PDC method involves the conversion of polymeric precursors (organometallic systems) into desired ceramic products with tailored chemical composition upon thermal treatment (pyrolysis), usually in inert environment [54–58]. Through this process, thermally and chemically stable ceramic products in complex forms that are usually difficult to achieve through conventional ceramic processing routes, such as powder sintering, can be easily prepared. This method leads to a simple and cost-efficient approach to fabricate ceramic components.

Since the precursors are usually in liquid state, the PDC technology gives an opportunity to print 3D components using slurry-based 3D techniques such as stereolithography (SLA). The manufacturing of 3D objects by SLA is based on the spatially controlled solidification of a liquid or semi-liquid resin systems by photo-polymerization such as the use of ultraviolet laser to selectively cure the liquid [59]. Such method was described by Eckel et al. [60] where preceramic monomers cured with ultraviolet light in a SLA printer were used to prepare 3D polymer structures of complex architecture and shape. The cured products were then pyrolyzed in argon environment to form ceramic components of silicon oxycarbide (SiOC). The mechanical result showed that the printed cellular SiOC material exhibited strength ten times higher than commercially available foams of similar density and survived at temperatures 1,700°C experiencing only surface oxidation (Figure 4.10).

4.4.2 PRINTED PDC SENSOR

3D printing of ceramics opens opportunities to fabricate complex-shaped, high-temperature, and environment-resistant ceramic components using the PDC technology for pressure sensing applications. This concept was displayed in Dr. Xu's lab (AI-M3) at North Carolina State University when a PDC wireless pressure sensor was fabricated via modified SLA process [61].

4.4.2.1 Printing Setup

An existing 3D printer (Lulzbot TAZ 4) was retrofitted to successfully print the preceramic polymer slurry. The major modification is on the extruder head where a syringe pump system is incorporated to serve as a plunger (Figure 4.11). A syringe is hooked to the extruder head and a slider controlled by a gear applies pressure on the extruder head in a translational direction. The printer is also incorporated with UV LED's surrounding the print stage to cure the polymer as it extrudes. The slurry is deposited layer by layer (20–100 μm) and the UV light shines upon the printed surface to expose a cross section for cross-linking. With this method, the cross-link efficiency depends on exposure parameters and can be increased by additional UV

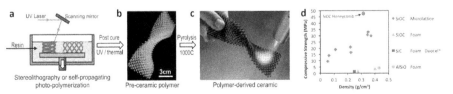

FIGURE 4.10 (a) Schematic of the SLA process. (b) Cured preceramic polymer after exposure to UV. (c) Ceramic product formed after pyrolysis. (d) Compressive strength of PDC–SiOC materials compared to ceramic foams [54].

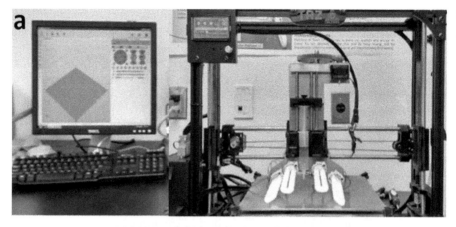

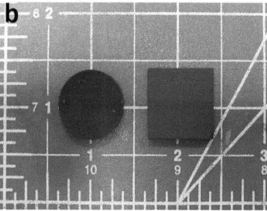

FIGURE 4.11   (a) Experimental setup for 3D printing PDC sensor. (b) Cylindrical and rectangular shaped PDC printed sensors.

exposure. A high cross-link efficiency is necessary to minimize the loss of small molecular gas species and crack initiation during pyrolysis. To minimize the possibility of crack initiation, the heating rate is controlled to be less than 20°C/minute. The print material consists of polymer-based ceramic precursor, carbon nanotubes (CNTs) as fillers, and UV curing agents.

4.4.2.2   Sensing Characterization of Pressure Sensor

The sensor was made using a rectangular prism SiC sample with dimensions of 18.06 mm × 18.06 mm × 2.6 mm. The sensor is modeled using design guidelines for a rectangular microstrip patch antenna (MPA). MPAs operate using a thin conductive strip above a ground plane separated by a dielectric material and designed to resonate at a specified frequency. The sample was then placed between the MPA mold. To interrogate the antenna, a National Instruments Vector Network Analyzer (VNA) PXIe-5630 (0.1–6 GHz) along with a wideband horn antenna (0.8–6 GHz) were used. The VNA applied a frequency sweep from 3 to 6 GHz in 0.0015 GHz steps. The horn is placed so that the antenna pattern is normal to the surface of the patch on the sensor. The sensor was placed in a Shimadzu AGS-J universal test machine (UTM). The sensor was positioned such that the patch faced toward the horn. The horn antenna was placed 6 inches away from the sensor. This distance was chosen due to the high reflective metal construction of the UTM. It is possible to increase this distance in low reflection environments or using a different interrogation antenna with a more focused beam pattern. A load was applied to the sensor in increments of 250 N. The resonance was recorded at each step until 4,500 N of force was loaded on the sensor (Figure 4.12).

4.4.2.3 Results

The resonant frequency of the sensor was recorded during increasing and decreasing loading. The resonant frequency of the sensor decreases as the load is applied. Figure 4.13a shows the S11 response of the sensor at various intervals as the load

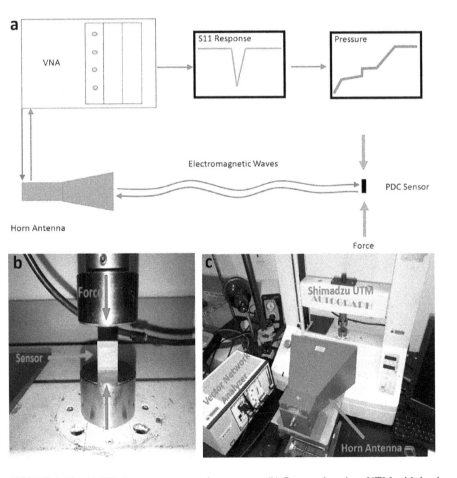

FIGURE 4.12 (a) Wireless pressure sensing process. (b) Sensor placed on UTM with load being applied to the thin edges. (c) Pressure sensor experimental setup.

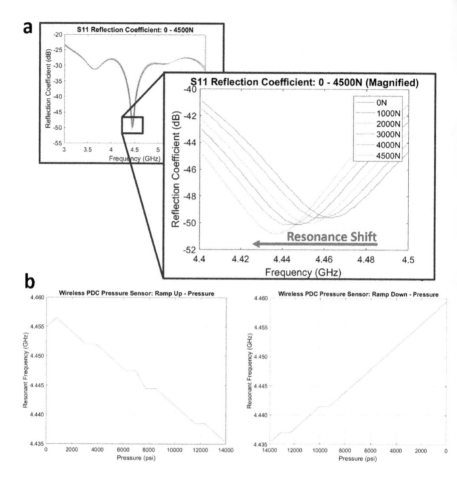

FIGURE 4.13 (a) As the pressure on the sensor increases, the resonance shifts to a lower frequency due to the increase in permittivity. (b) Ramp up Pressure vs. Frequency and Ramp down Pressure vs. Frequency.

is applied to the sensor. The pressure was determined using the area of the sensor that the load was applied and the known force applied using the load cell. From 0 to 13,800 psi, a roughly 20 MHz shift in resonance can be seen. After the loading ramp was applied, the sensor was still intact and had not cracked or deformed. From our results, we can see that the sensor's resonant frequency changes as a load is applied due to the permittivity property of the PDC.

4.5 CONCLUSIONS

The future of printed electronics continues to evolve to include more complex materials and designs. It is envisioned that other printed component technologies such as capacitors, antennas, and traces will be able to be designed and eventually seamlessly printed on a single machine for benefits such as weight reduction, cost savings for low production quantities of tailorable devices. Both monitoring and optimizing, if possible, pressures in extreme environments would be broadly useful for many defense applications to improve efficiency and safety.

ACKNOWLEDGMENTS

The authors would like to thank the Air Force Research Lab (Award No. FA8651-17-1-0005) for Chief Scientist Funds and the Air Force Research Lab Summer Faculty Fellowship Program at the Munitions Directorate. DISTRIBUTION A. Approved for public release, distribution unlimited. (AFRL-2023-3198).

REFERENCES

[1] David G. "Ceramic matrix composite (CMC) thermal protection systems (TPS) and hot structures for hypersonic vehicles", *15th AIAA International Space Planes and Hypersonic Systems and Technologies Conference, American Institute of Aeronautics and Astronautics*, 2008.

[2] Monteverde T, et al. Processing and properties of ultra-high temperature ceramics for space applications. *Materials Science and Engineering: A*. 2008;485(1–2):415–21.

[3] Fahrenholtz WG, et al. Refractory diborides of zirconium and hafnium. *Journal of American Ceramic Society*. 2007;90(5):1347–64.

[4] Klocke F, et al. Modern approaches for the production of ceramic components. *Journal of European Ceramic Society*. 1997;17:457–65.

[5] Wang X, et al. Diameter prediction of removal particles in Al2O3 ceramic laser cutting based on vapor-to-melt ratio. *Journal of Materials Processing Technology*. 2018;257:109–117.

[6] Lam CXF, et al. Scaffold development using 3D printing with a scratch-based polymer. *Materials Science & Engineering C*. 2002;20(1–2):49–56.

[7] ASTM F2792 – 12a. Standard terminology for additive manufacturing technologies. ASTM International.

[8] Khoshnevis B. Automated construction by contour crafting – related robotics and information technologies. *Journal of Automation in Construction – Special Issue*. 2004;13:5–19.

[9] Gebhardt A and Hotter J. *Additive Manufacturing: 3D Printing for Prototyping and Manufacturing, Chapter 1 – Basis, Definitions and Applications Levels*. Hanser, Munich, pp. 1–19, 2006. https://doi.org/10.3139/9781569905838.001

[10] Xu Y, Wu X, Guo X, Kong B, Zhang M, Qian X, Mi S, Sun W. The boom in 3D-printed sensor technology. *Sensors*. 2017 May;17(5):1166.

[11] García I, Zubia J, Durana G, Aldabaldetreku G, Illarramendi MA, Villatoro J. Optical fiber sensors for aircraft structural health monitoring. *Sensors*. Jun 2015;15(7):15494–519.

[12] Cheng H, Shao G, Ebadi S, Ren X, Harris K, Liu J, Xu C, An L, Gong X. Evanescent-mode-resonator-based and antenna-integrated wireless passive pressure sensors for harsh-environment applications. *Sensors and Actuators A: Physical*. 2014;220:22–33.

[13] Huang J, Zhou Z, Zhang D, Wei Q. A fiber Bragg grating pressure sensor and its application to pipeline leakage detection. *Advances in Mechanical Engineering*. Jan. 2015;5:590451.

[14] Pulliam WJ, Russler PM, Fielder RS. High-temperature high-bandwidth fiber optic MEMS pressure-sensor technology for turbine engine component testing. *Proceedings of SPIE - The International Society for Optical Engineering*. Feb. 2002;4578:229–238.

[15] Fu HY, et al. High pressure sensor based on photonic crystal fiber for downhole application. *Applied Optics.* 2010;49(14):2639–43.

[16] Fonseca MA, English JM, Von Arx M, Allen MG. Wireless micromachined ceramic
pressure sensor for high-temperature applications. *Journal of Microelectromechanical
Systems.* 2002 Nov 7;11(4):337–43.

[17] Baumann M, Ruther P, Paul O. Modeling and characterization of a CMOS sensor
with surface trenches for high-pressure applications. In *2011 IEEE 24th International
Conference on Micro Electro Mechanical Systems*, pp. 601–604, 2011.

[18] Blasquez G, Pons P, Boukabache A. Capabilities and limits of silicon pressure sensors.
Sensors and Actuators. 1989;17:387–403.

[19] Lee C, Tarbutton JA. Electric poling-assisted additive manufacturing process for PVDF
polymer-based piezoelectric device applications. *Smart Materials and Structures.* 2014
Aug 15;23(9):095044.

[20] Lin Y, Sodano HA. Concept and model of a piezoelectric structural fiber for multifunctional composites. *Composites Science and Technology.* 2008 Jun 1;68(7–8):1911–8.

[21] Baumann M, Ruther P, Paul O. Modeling and characterization of a CMOS sensor with
surface trenches for high-pressure applications. In *IEEE 24th International Conference
on MEMS*, pp. 601–604, 2011.

[22] Rakowski M, Segundo DS, Severi S, De Meyer K, Witvrouw A. "CMOS-integrated
poly-SiGe piezoresistive pressure sensor," *IEEE Electron Device Letters.*
2012;33(8):1204–6.

[23] Hanaoka Y, Fujimori T, Yamanaka K, Machida S, Takano H, Goto Y, Fukuda H.
One-dimensional-motion and pressure hybrid sensor fabricated and process-level-pack-
aged with CMOS back-end of line processes. In *Solid-State Sensors, Actuators and
Microsystems Conference*, pp. 684–687, 2009.

[24] Dongjo K, Sunho J, Sul L, Bong PK, Jooho M. "Direct writing of copper conductive
patterns by ink-jet printing," *Thin Solid Films.* 2007;515:7706–11.

[25] Suwanprateeb J, et al. Fabrication of bioactive hydroxyapatite/bis-GMA based composite via three dimensional printing. *Journal of Materials Science.* 2008;19(7):2637–45.

[26] Grimm T. Fused deposition modelling: a technology evaluation. *Time-Compression
Technology.* 2003;11:1–6.

[27] Bellini A, et al. New developments in fused deposition modeling of ceramics. *Rapid
Prototyping Journal.* 2005;11(4):214–220.

[28] Agarwala MK, et al. Structural ceramics by fused deposition of ceramics. In
International Solid Freeform Fabrication Symposium, 1995.

[29] Novakova-Marcincinova L. Application of fused modelling technology in 3D printing
rapid prototyping area. *Manufacturing and Industrial Engineering.* 2012;11(4):35–37.

[30] Venkataraman N, et al. Feedstock material property – process relationships in fused
deposition of ceramics (FDC). *Rapid Prototyping.* 2000;6(4):244–252.

[31] Travizky N, et al. Additive manufacturing of ceramic-based materials. *Advanced
Engineering Materials.* 2014;16(6):729–754.

[32] Shemelya C, Cedillos F, Aguilera E, Espalin D, Muse D, Wicker R, MacDonald E.
Encapsulated copper wire and copper mesh capacitive sensing for 3-D printing applications. *IEEE Sensors Journal.* 2014 Sep 16;15(2):1280–6.

[33] Shemelya C, Cedillos F, Aguilera E, Maestas E, Ramos J, Espalin D, Muse D, Wicker R,
MacDonald E. 3D printed capacitive sensors. In *SENSORS, 2013 IEEE*, pp. 1–4, 2013.

[34] Saari M, Xia B, Cox B, Krueger PS, Cohen AL, Richer E. Fabrication and analysis of a
composite 3D printed capacitive force sensor. *3D Printing and Additive Manufacturing.*
2016 Sep 1;3(3):136–41.

[35] Hehr A, Norfolk M, Wenning J, Sheridan J, Leser P, Leser P, Newman JA. Integrating
fiber optic strain sensors into metal using ultrasonic additive manufacturing. *JOM.* 2018
Mar;70(3):315–20.

[36] Chen Z, Li Z, Li J, Liu C, Lao C, Fu Y, Liu C, Li Y, Wang P, He Y. 3D printing of ceramics: a review. *Journal of the European Ceramic Society*. 2019 Apr 1;39(4):661–87.
[37] Zhang H, LeBlanc S. Processing parameters for selective laser sintering or melting of oxide ceramics. *Additive Manufacturing of High-performance Metals and Alloys-Modeling and Optimization*; Igor V. Shishkovsky Ed., IntechOpen Publisher, 2018 Jul 11:89–124.
[38] Hon KK, Gill TJ. Selective laser sintering of SiC/polyamide composites. *CIRP Annals*. 2003 Jan 1;52(1):173–6.
[39] Zhoum S. Absorption of some powder materials to YAG laser. *Science in China, Series A - Mathematics*. 2001;44(S1):489–94.
[40] Zhang T, Li Z, Li K, Yang X. flexible pressure sensors with wide linearity range and high sensitivity based on selective laser sintering 3D printing. *Advanced Materials Technologies*. 2019 Dec;4(12):1900679.
[41] Wei S, Zhang L, Li C, Tao S, Ding B, Zhu H, Xia S. Preparation of soft somatosensory-detecting materials via selective laser sintering. *Journal of Materials Chemistry C*. 2019;7(22):6786–94.
[42] Petrat T, Kersting R, Graf B, Rethmeier M. Embedding electronics into additive manufactured components using laser metal deposition and selective laser melting. *Procedia CIRP*. 2018 Jan 1;74:168–71.
[43] Le HP. Progress and trends in ink-jet printing technology. *Journal of Imaging Science and Technology*. 1998 Jan 1;42(1):49–62.
[44] Dong H, Carr WW, Morris JF. An experimental study of drop-on-demand drop formation. *Physics of Fluids*. 2006 Jul 6;18(7):072102.
[45] Kumar AV, Dutta A, Fay JE. Electrophotographic printing of part and binder powders. *Rapid Prototyping Journal*. 2004 Feb 1;10(1):7–13.
[46] Noguera R, Lejeune M, Chartier T. 3D fine scale ceramic components formed by ink-jet prototyping process. *Journal of the European Ceramic Society*. 2005 Jan 1;25(12):2055–9.
[47] Fromm JE. Numerical calculation of the fluid dynamics of drop-on-demand jets. *IBM Journal of Research and Development*. 1984 May;28(3):322–33.
[48] Bergeron V, Bonn D, Martin JY, Vovelle L. Controlling droplet deposition with polymer additives. *Nature*. 2000 Jun;405(6788):772–5.
[49] Lo LW, Shi H, Wan H, Xu Z, Tan X, Wang C. Inkjet-printed soft resistive pressure sensor patch for wearable electronics applications. *Advanced Materials Technologies*. 2020 Jan;5(1):1900717.
[50] Santiago CC, Randall-Posey C, Popa AA, Duggen L, Vuksanovich B, Cortes P, Macdonald E. 3D printed elastomeric lattices with embedded deformation sensing. *IEEE Access*. 2020 Feb 13;8:41394–402.
[51] Juhasz M, Tiedemann R, Dumstorff G, Walker J, Du Plessis A, Conner B, Lang W, MacDonald E. Hybrid directed energy deposition for fabricating metal structures with embedded sensors. *Additive Manufacturing*. 2020 Oct 1;35:101397.
[52] Cheng H, Shao G, Ebadi S, Ren X, Harris K, Liu J, Xu C, An L, Gong X. Evanescent-mode-resonator-based and antenna-integrated wireless passive pressure sensors for harsh-environment applications. *Sensors and Actuators A: Physical*. 2014 Dec 1;220:22–33.
[53] Colombo P, Mera G, Riedel R, Soraru GD. Polymer-derived ceramics: 40 years of research and innovation in advanced ceramics. *Journal of the American Ceramic Society*. 2010;93:1805–1837.
[54] Jia Y, Ajayi TD, Roberts Jr MA, Chung CC, Xu C. Ultrahigh-temperature ceramic-polymer-derived SiOC ceramic composites for high-performance electromagnetic interference shielding. *ACS Applied Materials & Interfaces*. 2020 Sep 23;12(41):46254–66.

[55] Jia Y, Chowdhury MA, Xu C. Complex impedance spectra of polymer-derived SiC annealed at ultrahigh temperature. *Journal of the American Ceramic Society*. 2020 Dec;103(12):6860–8.

[56] Jia Y, Ajayi TD, Xu C. Dielectric properties of polymer-derived ceramic reinforced with boron nitride nanotubes. *Journal of the American Ceramic Society*. 2020 Sep;103(10):5731–42.

[57] Jia Y, Chowdhury MA, Xu C. Electromagnetic property of polymer derived SiC-C solid solution formed at ultra-high temperature. *Carbon*. 2020 Jun 1;162:74–85.

[58] Jia Y, Chowdhury MA, Zhang D, Xu C. Wide-band tunable microwave-absorbing ceramic composites made of polymer-derived SiOC ceramic and in situ partially sur-face-oxidized ultra-high-temperature ceramics. *ACS Applied Materials & Interfaces*. 2019 Nov 14;11(49):45862–74.

[59] Arcaute K, Mann B, Wicker R. Stereolithography of spatially controlled multi-material bioactive poly (ethylene glycol) scaffolds. *Acta Biomaterialia*. 2010 Mar 1;6(3):1047–54.

[60] Eckel ZC, Zhou C, Martin JH, Jacobsen AJ, Carter WB, Schaedler TA. Additive manu-facturing of polymer-derived ceramics. *Science*. 2016 Jan 1;351(6268):58–62.

[61] Daniel J, Nguyen S, Chowdhury, MAR, Xu S, Xu C. Temperature and pressure wireless ceramic sensor (distance=0.5 meter) for extreme environment applications. *Sensors* 2021, 21:6648. https://doi.org/10.3390/s21196648

5 Metallization of 3D-Printed Devices

Nathan Lazarus
University of Delaware (formerly of US
Army Research Laboratory)

5.1 INTRODUCTION

Metals are everywhere in our day-to-day lives from electronics and computing to construction and structural components. With desirable properties ranging from high electrical and thermal conductivity [1] to strength and mechanical robustness [2], metals of all kinds are not just important but crucial to modern society. Metal 3D printing approaches like selective laser sintering (SLS) have therefore been among the most important technologies in additive manufacturing from the earliest days of the field [3]. However, metal-based additive manufacturing is limited in material selection [4], can result in porosity in the final print, and is far higher cost than other forms of 3D printing [5]. This reality has left metal 3D printing the purview of well-funded large companies and specialized research groups. Polymer-based 3D printers are far more common than metal ones on the market today [6] and are likely to remain so for the foreseeable future.

For creating robust and survivable electronics in harsh environments, bulk metallic conductors have been the default electrical traces and are the standard to which printed electronics are compared. In this chapter, one important avenue for achieving these metallic conductors and other parts in 3D printing will be discussed: metallization of parts printed using cheaper and more widely available plastic 3D printers. As additive manufacturing has matured, there has been increasing interest in hybrid processes combining 3D printing with complementary technologies in manufacturing and processing to enable exciting new capabilities and functionality [7]. Adding metals in particular can be used to provide advantages including improved mechanical properties [8], electrical [9] and thermal conductivity [10], corrosion resistance [11], lower surface roughness [10], and aesthetic properties [12]. During the next sections, commonly used approaches for integrating metal into or onto a 3D printed part will be reviewed, as well as the respective advantages and disadvantages of each. In a latter section of the chapter, a case study will then be discussed: the use of these metallization approaches to make an inductor, a common and fundamental circuit component whose need for high power handling capability and low resistance has made the device a crucial target of metallization in printed circuitry. The chapter finishes with conclusions and discussion of future research and growth areas in metallization of 3D printing.

DOI: 10.1201/9781003138945-5

5.2 APPROACHES FOR METALLIZATION

5.2.1 Electroplating

Electrochemical techniques are one of the most common methods for applying a surface coating commercially and unsurprisingly have become a preferred technique in 3D printing as well. Applying an electrical current through an electrolyte can be used to deposit metals on a surface, a process known as electroplating [13]. A voltage is applied between a positive electrode (the 'anode') and a negative electrode (the 'cathode') both in the electrolytic solution (Figure 5.1a). As a current develops between the electrodes, positive metal ions travel through the solution and neutralize on the cathode, causing a layer of metal to deposit on the cathode. Since an electrical current must be established to the electroplating target, only conductive surfaces can be electroplated directly.

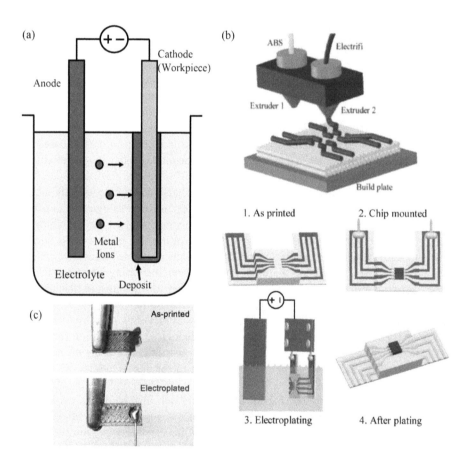

FIGURE 5.1 (a) Basics of electroplating. (b) Selective electroplating process for FFF. Reproduced with permission [20]. Copyright 2019, John Wiley and Sons. (c) Soldering to plated copper films. Reproduced with permission [21]. Copyright 2019, Elsevier.

One of the biggest challenges in using electroplating for depositing metals on parts printed using polymer-based additive manufacturing is the need for a conductive cathode. Most 3D-printed plastics are not electrical conductors and therefore require additional processing to render the surface conductive for electroplating. One simple option is to take the final 3D printed part from a process like fused filament fabrication (FFF) and use a paintbrush or spray coater to add a layer of conductive paint on the surface [14]. The conductive surface layer is then used to deposit a final layer of copper or other electroplated metal on the final part. This technique does, however, have the limitation of requiring manual painting of individual parts and is not easily scaled to more complex patterns or geometries.

Alternatively, a non-conductor can also undergo an initial blanket coating step to create a seed layer on the surface. In one common approach, seed clusters of conductive particles such as palladium are first deposited onto the non-conducting surface in an initial metallization [15]. As current is applied to an attached cathode electrode, plated metal rapidly bridges between the individual seed clusters to form a continuous film to plate the surface. This approach is effective, but non-selective, resulting in a continuous coating rather than the independent traces needed for electronics.

There are however advantages to being able to direct print the conductive regions within the part, rather than requiring additional patterning steps for metallization. Conductive composite materials have become increasingly common within the 3D printing community, and many options are now commercially available [16]. In FFF, for example, thermoplastic filaments loaded with conductive particulate such as carbon black or metal particles have been developed. This particle loading allows parts to be printed where some or all of the part is conductive, albeit with orders of magnitude higher resistance than traditional bulk metal conductors like copper. The particles serve as conductive paths through the otherwise non-conducting thermoplastic. Conductive composites can therefore be plated much like other conductors, with the plating tending to anchor on and nucleate from particles exposed on the surface [17].

Relying on this approach, electroplating has been demonstrated directly on 3D printed thermoplastic composites made using FFF[18]. Using a FFF printer with dual extruder heads (Figure 5.1b), parts were printed combining both a conductive thermoplastic composite and a non-conducting thermoplastic filament. The use of a dual extruder allowed individual regions to be selectively electroplated, allowing different thicknesses or materials to be plated in close proximity in the same part. This approach can be performed both on carbon/graphene loaded composites [18,19] as well as on filaments based on metal particles [20,21]. Since electroplating is capable of depositing relatively thick layers, on the order of hundreds of micrometers, soldering is possible to the resulting deposits, where unplated composites melt during the soldering process (Figure 5.1c). The ability to solder is an important enabling development, allowing more traditional electronics assembly to be performed on the part with minimal modification. A similar electroplating approach combining non-conducting FFF parts with direct ink writing (DIW) of silver inks has also been shown for selective electroplating of 3D printed parts [22], showing that this method is not limited to FFF.

For metallization of 3D printed parts, electroplating based on an acidic copper sulfate chemistry is the most widely used, and it was the choice for the demonstrations above [18–22]. This plating chemistry deposits metallic copper, largely for use as a low resistance electrical conductor. A typical formulation consists of 150–250 g/L hydrated copper sulfate ($CuSO_4 \cdot 5H_2O$) in approximately 40–60 g/L sulfuric acid, with plating performed at or near room temperature [13]. Plating rates are on the order of tens of micrometers per hour [18]. Standard electroplating baths result in a relatively coarse grain structure with grain size on the order of a micrometer or larger and correspondingly high surface roughness; organic additives known as levelers can be added to the bath to reduce the grain size and obtain a smoother final surface finish [23]. Since surface deposits on 3D printed parts have relatively high roughness even before plating, a similar approach has been investigated on 3D printed composite filament traces with demonstrated roughness reductions by a factor of three compared to the initial printed trace and a factor of two better than electroplating alone [21].

For the purposes of printed electronics, electroplating has several crucial advantages. The technology is mature and widely used commercially, making electroplating well understood and easily available at low cost. Layers up to hundreds of micrometers thick are possible in a few hours of plating, and the resulting deposits are both solderable and comparable to bulk metal. In terms of disadvantages, the most notable is the requirement to make electrical contact with the area to be plated. Electroplating is capable of plating independent regions of a part, but circuits with a large number of isolated traces are difficult to plate completely. Plating a circuit with independent electrodes simultaneously has been demonstrated using a breakaway region that is removed after plating [20], but this requirement complicates the geometry and may not be possible for all cases. Electroplating also requires exposing the part to harsh chemistries, often strong acids [24], albeit chemistries that typically do not react significantly with most 3D printed polymers.

5.2.2 ELECTROLESS PLATING

One of the biggest limitations of electroplating is the need to make electrical contact. A current must be established to any individual conductor intended to be plated, making the plating of a system with numerous independent conductors relatively complicated. Electroplating is also most easily performed on the outermost conductive surface of a part, and maintaining uniform layers of deposition on a complex part with numerous trenches or other high aspect features is very difficult [25]. In the 1940s, researchers found that plating could be initiated from certain plating baths onto a surface with a suitable catalyst, using the reaction between the bath and the surface to trigger and maintain the deposition in a process known as electroless or autocatalytic plating [26]. Electroless plating relies on the presence of a reducing agent to contribute electrons to metal ions in the solution, which then react and deposit on the exposed catalytic surface (Figure 5.2a) [27]; the charge from the reducing agent eliminates the need for an external current. Since electroless plating does not rely on an applied current, the process is highly conformal and is capable of both plating deep into high aspect ratio features and onto numerous independent regions on a target simultaneously [28]. However, electroless plating is significantly

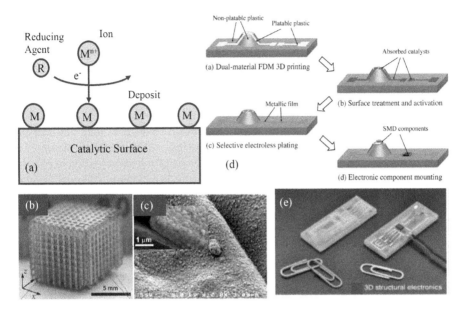

FIGURE 5.2 (a) Electroless plating process, example part (b) after electroless plating and (c) image of microstructure. Reprinted with permission [34]. Copyright 2014, American Chemical Society. (d) Example selective electroless plating process and (e) resulting part. Reproduced with permission [38]. Copyright 2018, John Wiley and Sons.

slower than electroplating [27], with typical rates on the order of a few micrometers an hour [29], and as a result, it is often followed by an electroplating step to build additional thickness [30], as in the 3D printed antenna in Ref. [31].

As with electroplating, electroless plating on a polymer 3D part typically requires significant sample preparation to render the surface suitably catalytic for plating. Typically, this process consists of a series of steps including an etch to roughen the surface for better adhesion, neutralization of the etchant on the surface, followed by an activation step to introduce a precious metal catalyst on the surface [32]. At this point, the part can be placed into an electroless bath for plating. Using this approach, there have now been demonstrations of electroless plating of parts made using a number of different polymer-based 3D printing approaches including stereolithography (SLA) (Figure 5.2b and c) [33,34], FFF [35], multi-jet modeling (MJM) [36], and two photon polymerization [37].

Electroless plating is typically a blanket process, giving uniform coatings over the entire exposed surface of a part. Selective plating of the independent conductors needed for applications such as 3D-printed electronics can however be achieved through several different techniques. One method is to use a part consisting of multiple different materials, such as an FFF part printed using dual extrusion of different thermoplastics [38]. If only one of the thermoplastics reacts with the activation solution for applying the catalyst, electroless plating can be controlled to only occur on that plastic, resulting in a selective deposition process (Figure 5.2d and e). Alternatively, catalyst can also be patterned either lithographically [39] or through activation with a laser system ('laser direct structuring') [40] to obtain independent regions for plating. Recently, there has also been a demonstration of selective printing using filament loaded with catalyst, with dual extrusion of a conventional thermoplastic with a second head to selectively deposit plateable regions [41]. As with electroplating directly on conductive filaments, this technique is a particularly promising avenue toward allowing complex structures to be directly printed and plated.

For electroless plating of 3D printed parts, the selective nickel plating process in [38] is a representative example. First, a chemical etch based on a chrome/sulfuric etchant (400 g/L chromium trioxide plus 400 g/L sulfuric acid in DI water) is used to roughen the surface of ABS by removing one of the two constituent phases. Activation is then performed using commercial 1.5 g/L palladium colloidal catalyst (Pd/Sn) in a hydrochloric acid/salt solution (280 mL/L HCl and 50 g/L NaCl). A commercial accelerator solution is then used to neutralize the Pd particles and wash the catalyst from the regions of other (non-ABS) polymer to achieve selective plating. Finally, the part is electrolessly plated in a nickel plating solution consisting of a nickel salt ($NiSO_4 \cdot 6H_2O$, 30 mL/L), a reducing agent ($NaH_2PO_2 \cdot H_2O$), and a complexing agent ($Na_3C_6H_5O_7 \cdot 2H_2O$) in DI water. A complexing agent serves to prevent precipitation of metal salts from the bath to enable plating [42].

As with electroplating, electroless plating is widely used and commercial solutions are readily available. The process also has the clear advantages of eliminating the need for a second electrode as well as the ability to plate multiple independent regions simultaneously and to create conformal coatings, including deep into high aspect features. As with electroplating, the resulting deposits are also comparable to bulk metals. However, electroless plating is much slower than electroplating and films thicker than a few micrometers are time consuming. The process also requires special processing or materials to achieve selectivity, and activation solutions tend to be expensive due to their precious metal constituents.

5.2.3   LIQUID METALS

With the growth of microfluidics for chemical and biological applications, there has been progress toward developing and refining techniques for additive manufacturing of fluidic channels and other systems [43]. These techniques have opened up an alternative approach toward metallizing a 3D printed part: flowing a conductor into printed channels. In addition to the use of more conventional conductors such as silver inks [44], there have been a number of demonstrations with metals that are liquid at room temperature such as certain gallium alloys [45]. Proposed as an alternative to other, more toxic liquid metals such as mercury, the liquid gallium alloys are now widely used in the areas of tunable and stretchable/wearable electronics due to their ability to easily flow and reconfigure to match their surroundings. This approach to metallization can therefore be combined with 3D printed elastomers to create fully stretchable systems [46].

The alloys eutectic gallium–indium (eGaIn) and galinstan (an alloy of gallium, indium, and tin) are non-toxic and have an electrical conductivity of approximately 3.4×10^6 S/m [47], roughly 17 times lower than copper. When exposed to air, these alloys also form a nanometer-scale oxide layer that resists motion until a threshold is reached and the oxide is broken [48]. This property prevents undesired leakage and also has enabled the ability to create free-standing structures using the liquid metal alone (Figure 5.3a) [49].

For the use of 3D printing to create microfluidics for liquid metal conductors, several approaches have been investigated. Within the stretchable electronics community, soft silicones are widely used for their biocompatibility and ability to match the mechanical performance of human skin and tissue [50], but these materials are relatively difficult to 3D print in many conventional processes [51,52]. However, 3D printing is widely used to create molds to define channels in silicone, which can then

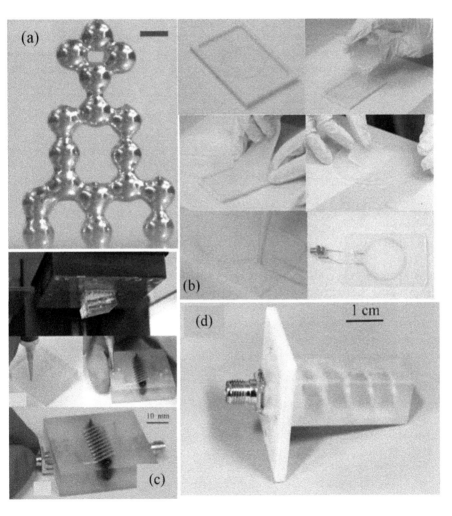

FIGURE 5.3   (a) Stacked 3D printed liquid metal droplets. Reproduced with permission [49]. Copyright John Wiley and Sons 2013. (b) Molding liquid metal devices from 3D printed molds. (c) Direct printing of channels for liquid metals in SLA. Reproduced with permission [55]. Copyright 2019, Elsevier. (d) Helical antenna using vacuum filled channels. Reproduced with permission [56]. Copyright 2017, Elsevier.

be followed by injection of liquid metal (Figure 5.3b) [53,54]. More recently, there have been major advances in direct printing of fluidic channels for liquid metal fill (Figure 5.3c and d). Liquid gallium alloys have been injected into parts made using FFF [46], SLA [55], and MJM [56].

Liquid metals and other liquid injection-based approaches are most valuable in being relatively material independent; any printed channel can be filled with a liquid conductor to form a circuit, and there is no need for the 3D printed part to be conductive or have specialized surface properties. Both this material independence and the ability to flow even after injection are what makes this option particularly promising in the stretchable and reconfigurable electronics spaces. However, the electrical resistivity is also substantially higher than copper, so this approach is most suitable for specialized applications that are not possible with traditional materials. Although liquid metals can create effective electrical connections between wires, they do not create a mechanical joint similar to a soldered joint since they do not solidify; if a mechanically robust joint is needed, an alternative conductor would be required.

5.2.4 SINTERING

Similar to the liquid metals, conductive inks are another approach for allowing deposition of metals in liquid form. Silver flakes or other conductive particles in a liquid can be directly printed to create electrical traces with low resistivity [57]. This approach has been adapted repeatedly for 3D printed conductors, typically using DIW [58] or aerosol jet printing [59] to deposit the conductive inks. However, as printed, most conductive inks have resistivity orders of magnitude larger than bulk silver or other conductors due to the presence of organic molecules to stabilize printing [60]. Such inks typically require a sintering step at elevated temperature to achieve optimal electrical properties. Sintering is defined as heating a powder below the powder's melting point to cause the particles to coalesce and bond together [61]. This bonding of the individual particles together results in a more continuous metal film and therefore significantly lower electrical resistance. The needed temperature range varies substantially from near room temperature to hundreds of degrees Celsius depending on the properties of the individual silver ink used.

Typical silver inks consist of silver nanoparticles or nanowires with colloidal stabilizers to minimize clogging and require temperatures on the order of 100°C–200°C for sintering [58,62]. Inkjet printing requires a relatively low viscosity (from one centipoise [63] to a few hundred centipoise for newer techniques [64], while aerosol jet is capable of printing inks with viscosities between 1 and 1,000 cP [65]. A typical commercial silver ink for inkjet or aerosol jet printing with solids content on the order of 50% by weight has a viscosity of 50–100 cP [66]. For extrusion-based DIW, higher viscosities are typical (10^6–10^8 cP) [67], allowing solids content of 90% or higher [68]. More recently, new non-particle reactive silver inks have also become available based on a silver salt such as silver acetate [69,70], with notable advantages including lower temperature sintering (<100°C) and the elimination of particles that could cause clogging in a nozzle.

While silver inks are widely used, the cost of silver metal has made these types of ink relatively expensive, and copper-based inks have also been explored as a low-cost alternative [71]. The major challenge has been oxidation; copper oxidizes easily to form non-conducting copper oxides at the temperatures commonly used for sintering of conductive inks [72,73]. To avoid oxidation, particles in the ink typically must be coated in organic [71,74,75], coatings that protect the metal from oxygen exposure.

The most straightforward sintering approach, heating in an oven or other similar system [76], is widely used but has sizeable drawbacks in additive manufacturing. Many 3D printed polymers have glass transition temperatures of 100°C or less, below the optimal temperatures of many conductive inks. Effective use of oven sintering therefore requires either an ink with a very low sintering temperature or the use of a more specialized polymer capable of surviving higher temperatures. As a result, a variety of alternatives have been investigated to allow more robust sintering even on low temperature polymers.

Since metal particles absorb light differently than neighboring polymers, the use of light-based approaches has been particularly common. A flash of white light from a xenon lamp can be used to selectively heat conductive traces above the temperatures needed for sintering due to absorption into the contained particulate, while neighboring polymer regions remain at lower temperatures. Intense pulsed light (IPL) sintering [77], as this approach is called, allows for rapid sintering of numerous traces across a printed part. In applications where selective sintering is desired, focused laser heating has also been demonstrated for sintering conductive inks [78]. Metal particles are first deposited on the substrate using a variety of approaches such as inkjet or spin coating; the deposition is then followed by a brief pre-heating and subsequent metallization by focused laser pulses to create the final pattern. Selective sintering can also be achieved through resistive heating of individual traces through an applied electrical current [79]. In Ref. [79], nanoparticles were found to coalesce at power levels of between 0.1 and 10 mW/μm^3, and carbonization of the organic shells was also found to occur and to contribute to the drop in electrical conductivity.

Each of these sintering strategies has been applied to conductive inks on 3D printed parts. For instance, in Ref. [62], six sintering methods (oven baking, IPL, laser heating, resistive heating, microwave heating, and atmospheric plasma) were investigated for inkjet-printed silver nanoparticle inks on acrylate polymer parts 3D printed using MJM. Example traces for four of these approaches are shown in Figure 5.4a–l. Oven heating was found to give the most consistent and lowest resistivity films, while the varying heat distribution resulting from the other techniques results in more porous and less conductive films. Since many of the approaches were developed in the context of printable electronics for flexible printed circuits, adapting these techniques to fully 3D shapes can also be problematic; the commercial systems for IPL, for instance, focus the intense light pulse in a 2D plane resulting in more limited sintering for more fully 3D designs [62]. Using IPL on the thicker traces printed by DIW also can require longer or repeated exposures to obtain reasonable sintering [80].

Sintering has also been demonstrated for creating metal parts using FFF with composite filaments. In the 1990s, a technology known as fused deposition of metals (FDMet) was developed using polymer filaments heavily loaded with metal particles; after printing, thermal or chemical removal of the polymer binder followed by sintering is used to create the final metal part [81]. Our group has also recently

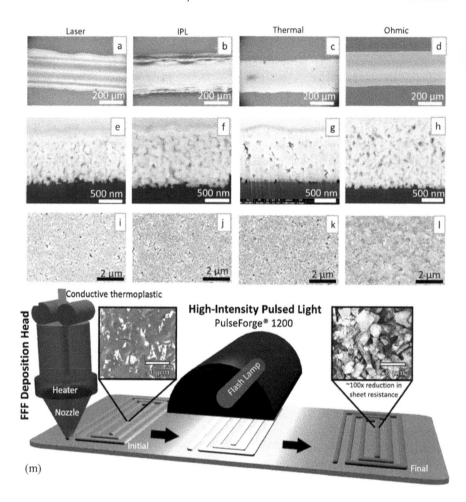

FIGURE 5.4 (a–l) Sintering silver inks using different approaches. Reproduced under the terms of the CC-BY Attribution 3.0 license [62]. Copyright the authors 2017. (m) Flash ablation metallization. Reprinted with permission [82]. Copyright 2020, Elsevier.

investigated the use of IPL exposure on conductive thermoplastic composites. Although FFF-based 3D printing is both cheaper and more widely available than printing of silver inks, the available conducive thermoplastic composites have not been competitive in performance. The best performing composite filaments available commercially are based on metal particles within a thermoplastic matrix and have a resistivity of 0.014 Ω·cm [16], compared with resistivity as printed on the order of 1×10^{-4} Ω·cm or as low as bulk silver (1.6×10^{-6}) after annealing to moderate temperatures for printed silver inks [70]. While an improvement over earlier conductive filaments, these filaments remain too resistive for many applications.

Since the best performing conductive filaments are based on metal particles within a polymer matrix, IPL was also investigated for conductors printed using FFF (Figure 5.4m) [82]. Unlike silver inks, the filament used does not drop in resistivity with simple heating in an oven, likely due to the high concentration of thermoplastic; instead, the resistivity increases dramatically. However, at very high powers, the surface polymer ablates away completely resulting in a metal-rich surface layer in a process called Flash Ablation Metallization (FAM). This metal-rich surface layer leads to a reduction in resistivity by up to two orders of magnitude compared with the initial composite trace.

Silver inks and sintering have the clear advantage of allowing direct printing of metals in desired locations using extrusion or other relatively cheap polymer 3D printing techniques. Conductivity can also be brought close to bulk metal properties with appropriate inks and sintering conditions. On the other hand, the need for high temperatures can limit the use of these techniques to higher temperature polymer parts for optimal performance, or require more specialized and expensive processing than a simple oven bake. This approach also typically results in relatively porous conductors which can limit long-term reliability and power handling [83,84].

5.2.5 SPUTTERING AND EVAPORATION

The microelectronics industry is reliant on depositing very well controlled layers on silicon wafers for the purposes of making complex circuitry down to the nanometer scale. These requirements have led to the development of highly specialized metal deposition tools that researchers have experimented with for the purposes of metallization of 3D printed parts. For metal deposition, the most commonly used technology in the semiconductor industry is known as sputtering [85]. Under vacuum, a substrate is placed across a gap from a block of solid metal known as a sputtering target [86]. An inert gas, typically argon, is then allowed to flow into the chamber [87]. After the argon is ionized in a plasma, the ions are accelerated against the sputtering target through a bias electric field [88], knocking loose atoms from the target that then re-deposit on the substrate to form a polycrystalline metal film on the surface [89].

Sputtering is only one of several processes for transferring metal particles physically from a source onto the substrate of interest, known collectively as 'physical vapor deposition' [87]. Sputtering relies on physical bombardment to move particles off the target; however, heating a metal sufficiently also can cause vaporization, a process known as evaporation. In evaporation-based deposition, either resistive heating or an electron beam is used to heat a metal source beyond the point where the metal begins to evaporate; the metal particles then condense on exposed surfaces within the chamber to form a metal film [90]. Sputtering is a relatively conformal deposition technique with relatively good step coverage [91], meaning the deposits will coat both the top and sides of a vertical step, while evaporation is highly directional with minimal step coverage [92].

Due to the cost of the equipment involved, sputtering and other vacuum deposition techniques have remained a relatively niche approach toward metallization in additive manufacturing. Use of the technique within 3D printing has been the most prominent for specialized RF components such as antennas [93] and transmission lines [94], where the need for smooth, well controlled surfaces is particularly important [95]. This metallization approach has also been demonstrated for improving more generic properties of a 3D printed part like durability [96], although this technique would be very costly compared with alternatives like electro- and electroless plating since this application does not rely as heavily on the advantageous properties of vacuum deposition. As with other metal deposition approaches, adhesion to the plastic substrate can be challenging for sputtering metals like copper that are desired for their electrical properties. When sputtering on difficult substrates, an additional layer of a metal known as an adhesion layer is typically sputtered prior to the metal of interest [97]. This adhesion layer, commonly titanium or chromium [98], is chosen to adhere well to most substrates as well as to the follow-on metal layer to be deposited. In [94], for instance, a thin layer of titanium (60 nm) was added to improve adhesion to the 3D printed surface (a proprietary polymer known as VeroWhitePlus) before the addition of a thicker layer (200 nm) of copper for copper's electrical properties. Evaporation for metal deposition on 3D printed parts has also been demonstrated, for instance, copper on a chromium adhesion layer on an FFF ABS part in Ref. [99].

Vacuum deposition technologies are capable of depositing extremely well controlled metal layers with bulk properties, and these characteristics have made them a promising option in certain specialized applications such as RF devices. However, the high cost and limited availability of these specialized systems limits widespread use of this approach, and this technique is likely to remain a niche within additive manufacturing. Vacuum deposition approaches also can result in substantial heating, and there have been demonstrations showing some lower temperature 3D printed parts can be damaged during the process [100].

5.3 APPLICATION EXAMPLE: INDUCTORS AND WIRELESS POWER

For electronics, metallization is most crucial for applications that require either low resistance or improved current handling. Many circuits in areas of interest to the 3D printed electronics community like sensing, control, and communication can be designed to be relatively insensitive to conductor resistivity. A poor conductor might affect speed or performance, but basic functionality can likely still be maintained. The same cannot, however, be said of power components; important metrics like efficiency are unavoidably related to conductor performance. A higher resistance results in energy being burned as heat, not only wasted from the system's perspective but also risking damage to the trace or any low melting point polymers nearby. Obtaining electrical resistivity as close as possible to metals like copper is therefore crucial to creating useful power systems using 3D printing.

When a current is passed through a wire, a magnetic field is generated (Figure 5.5a), and the resulting fields cannot change instantly, allowing these fields to be used for energy storage in a device known as an inductor. Since neighboring wires have mutual electromagnetic coupling between them that allows for higher energy storage density, most inductors consist of coils of wire, with common examples including a coil with all turns in the same plane (a planar coil) and a coil where the turns instead follow a cylindrical geometry (a solenoid) (Figure 5.5b). The motivations for wanting to print an inductor or other electronic passive component include faster product development, improved supply chain efficiency, and the ability to create or integrate into more complex, 3D structures [101]. As one of the fundamental circuit passive components, there are a wide variety of applications, but in 3D printing, most

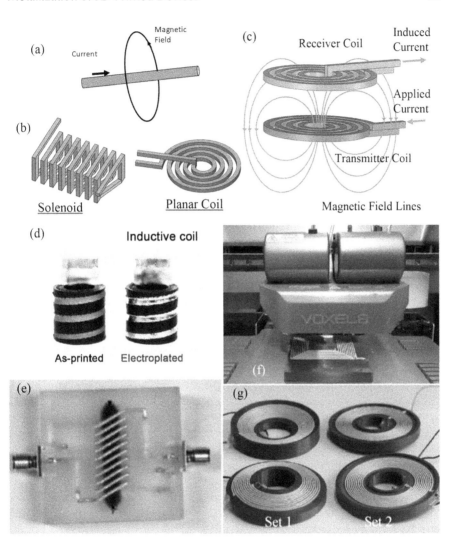

FIGURE 5.5 (a) Magnetic field generation, (b) common inductor designs, (c) inductive wireless power and 3D printed inductors made using (d) electroplating. Reprinted with permission [21]. Copyright Elsevier 2019. (e) Liquid metal fill. Reprinted with permission [55]. Copyright Elsevier 2019. (f and g) DIW silver ink. Reproduced under the terms of the CC-BY Attribution 4.0 license [104]. Copyright the authors, 2019.

common inductors are used for power conversion [102,103], wireless power transfer [104], and electromagnetic field generation [105] for actuation. Power converters are a family of circuits that shift energy between power waveforms, including from one voltage level to another or between frequencies, as well as from AC to DC voltage and back [106], with inductors commonly used as energy storage elements. For wireless power transfer, magnetic fields generated by one coil are used to induce a current in a second inductor for transmitting power (Figure 5.5c) [107]. For actuation, inductor coils are used to move a magnet, a technology widely used for motors [108] and speakers [109]; in the inverse, a moving magnet can also be used to induce a current in a coil for sensing or energy harvesting [110].

Inductors are fundamentally an energy storage element. As a result, one of the key specifications of an inductor is a measure of how efficiently the inductor stores energy, the quality factor Q. The parameter Q is defined as the ratio of the reactive energy stored to the average resistive losses, or:

$$Q = \frac{2\pi f}{R} L \qquad (5.1)$$

where L is the inductance, f is the frequency of operation, and R is the series resistance of the inductor [111]. Q is therefore frequency dependent, and inductors in general store energy better at higher frequencies. However, at a frequency known as the self-resonant frequency, the capacitance between the turns resonates with the inductance and the inductor begins to turn capacitive [112]; this result means that the inductor Q rises for low frequencies, then peaks near self-resonance before dropping off. The highest Q results when the inductance L is as large as possible relative to the resistance R, driving the need for high-quality metal conductors to maximize energy storage performance.

Since an inductor is a simple coil of wire, any 3D printable conductor can be used to print one, although not necessarily one with good performance. As an example, conductive thermoplastics are typically several orders of magnitude more resistive than a copper trace, but there have been demonstrations of inductive-based wireless power systems capable of transmitting a small amount of power, enough to light up an LED [16]. While not specifically reported in [16], the quality factor is unlikely to be competitive with traditional wireless power systems. The inductor is likely to have a quality factor that is smaller than that of a comparable copper wire. However, there are applications where power efficiency is not a significant concern (where, for instance, ample amounts of access to the power grid are available and small amounts of power are needed to perform a function), and such a device may still be useful.

For example, 3D printed inductors can be used in applications where more traditional bulk metals are simply unsuitable. The field of stretchable electronics is one important example [113]. While conventional metals like copper are rigid and not easily stretched without careful thinning and structuring, liquid metal traces can survive deformations of tens or hundreds of percent without damage [114]. Since these types of systems are intended to survive repeated strains, making physical contact to conventional rigid wiring is particularly difficult; as a result, the use of wireless transmission for power and communication has been used in a number of stretchable systems [115–117]. New 3D printing technologies for printing elastomers have opened up the ability to directly print fluidic channels into soft systems [118], allowing the fabrication of liquid metal inductors and other devices. Liquid metal inductors have been fabricated both using casting from 3D printed molds [53,54], as well as through direct printing of channels (Figure 5.5e) [55] followed by fill with liquid metal, including the highest Q stretchable inductors ever reported (as high as 174) [119].

The ability to create larger cross sections therefore enables liquid metals in 3D printed devices to be competitive or superior to other stretchable devices despite not being as conductive as traditional conductors.

Directly competing with conventional inductor technology typically has required the use of metallization using more traditional metals. Casting of sterling silver based on a 3D printed inductor has been performed successfully, although the resulting traces were found to have only a third the conductivity of bulk copper [102]. A similar conductivity drop was also found for devices printed using DIW of silver inks that were then sintered at temperatures as high as 250°C [120], although this conductivity is still six times higher than that reported for inductors with unsintered silver ink (Figure 5.5f and g) [104]. The inductor in Ref. [120] had a quality factor of approximately 10 at 1 MHz, with total inductance ~150 nH and a DC resistance of 0.1 Ω. This quality factor also benefited substantially from the inclusion of a printed magnetic core with permeability of 23; inclusion of a high permeability magnetic material results in higher energy storage and therefore higher inductance.

For the purposes of achieving bulk metal performance, however, copper plating (both electroless and electroplating) has remained the state of the art, since the performance of the resulting conductors is comparable to bulk metal. Electroless plating is most effective for higher frequency inductors where the skin effect limits the flow of electrical currents to the conductor surface [103], and there have been 3D printed inductors with electroless plated copper designed for high frequency ranges [9]. The skin effect is a phenomenon in high frequency conductors where alternating electrical currents within the conductor create eddy currents that limit current density to the surface of a conductor. As a result, for high frequency conductors only the surface of a conductor is usable for carrying electrical current, and only a limited thickness of metal is needed. Since electroplating is capable of significantly thicker films, up to hundreds of micrometers, electroplating is better suited for inductors aimed at lower frequencies. Several electroplated inductors have also been demonstrated, based on electroplating on devices printed using conductive composites in FFF [18,21]. Using this approach, very low resistances are possible; for instance, the three turn composite inductor in Ref. [21] (Figure 5.5d) reached resistances of roughly 30 mΩ, compared with tens of Ohms prior to plating, and a quality factor of 13.7 at 1 MHz. At higher frequencies, the quality factor can be even higher; the inductor in Ref. [103], based on an SLA inductor plated using a commercial plating service (full process details unclear, but likely electroless plating followed by electroplating to make a thickness of approximately 50 μm of copper), had a quality factor as high as 94 at approximately 40 MHz.

While 3D printed inductors can be made in any process capable of creating a meaningful conductor, metallization is crucial to performance, both in terms of electrical resistance and lifetime. Of the metallization approaches available, liquid metals and sintered silver inks have typically resulted in resistivity substantially lower than bulk metals; these options are therefore most useful in niche applications where the material or geometry advantages outweigh the loss in performance, with stretchable electronics being one notable example. Electro- and electroless plating are the most capable of depositing high-performance conductors and are therefore preferred for applications that require performance. While vacuum deposition approaches are capable of depositing high-performance conductors, the cost and geometry requirements have limited their use in 3D printing in general, and the author is unaware of examples of their use for inductor fabrication.

5.4 CONCLUSION

From sintering to plating, metallization is a powerful approach to adding new materials and capabilities to a 3D printed part. Due to the high costs associated with metal 3D printing, using post-processing to add metals to a polymer part has the potential to make metals far more widely available in additive manufacturing. The resulting metal films also tend to be both denser and closer to bulk properties, improving performance and long-term reliability. While complex structural parts will continue to be done using metal 3D printing, metallization is likely to develop into an important complementary technology for applications like printed electronics that benefit from the higher resolution and improved material properties that metallization can provide.

As this area grows, there remain significant fertile research areas that remain open for future advancement. One important area is in broadening the available material set. Metallization efforts have predominantly focused on a handful of specific metals, most notably nickel for mechanical properties and silver and copper for electrical ones. Most of the basic deposition technologies have been demonstrated for a far broader set of possible metals and metal alloys with useful properties such as magnetic performance or a low melting point for soldering. Integrating these alternative materials could provide significant improvements in functionality with only limited further development.

Much of the challenge inherent in metallizing a 3D printed part is in achieving a strong interfacial bond, and developing a more detailed understanding of this interface is also an important area of future research. Metallization research has been using conventional and well established metallization technologies adapted from other industries, and there has not been extensive work in characterizing the strength and nature of the interface for a given technique. Since the layered nature of 3D printing results in specific, repeated surface features and roughness, understanding the unique properties of the resulting metallization process might lead to new advances in improved surface finishing. While the focus of this chapter has been on deposition approaches, there have also been recent advances investigating embedding copper wires and meshes that have shown significant promise [121,122]. Metal wires are embedded during the printing process, allowing the wires to be placed within the body of the printed parts. This approach has the advantage of direct use of bulk conductors, albeit with geometry limitations; combining other metal deposition approaches with metal embedding could be particularly advantageous.

Perhaps the biggest likely growth area, however, is in broadening the application space. Since many of these metallization approaches are relatively new, having been demonstrated only in the past few years, metallization is most widely used in only a handful of specific application areas like inductors, antennas, and circuitry. Most notably, there is a potential to use metallization to directly compete with the applications currently dominated by metal 3D printing; metallization is both cheaper and more widely available, and there is significant promise in being able to use technologies like electroforming [123] to directly make similar parts, which could revolutionize the industry.

ACKNOWLEDGMENTS

This work was performed at and funded by US Army Research Laboratory through internal research funding. The views and conclusions contained in this chapter are those of the author and should not be interpreted as representing the official policies, either expressed or implied, of the Army Research Laboratory or the U.S. Government. The U.S. Government is authorized to reproduce and distribute reprints for Government purposes notwithstanding any copyright notation herein.

REFERENCES

[1] R. Resta, "Why are insulators insulating and metals conducting?" *J. Phys: Condens. Matter*, vol. 14, 2002, pp. R625–656, https://doi.org/10.1088/0953-8984/14/20/201
[2] K. Lu, "The future of metals," *Science*, vol. 328, 2010, pp. 319–320, https://doi.org/10.1126/science.1185866
[3] S. H. Huang, P. Liu, A. Mokasdar, and L. Hou, "Additive manufacturing and its societal impact: a literature review," *Int. J. Adv. Manuf. Technol.*, vol. 67, 2013, pp. 1191–1203, https://doi.org/10.1007/s00170-012-4558-5
[4] W. E. Frazier, "Metal additive manufacturing: a review," *J. Mater. Eng. Perform.*, vol. 23, 2014, pp. 1917–1928, https://doi.org/.1007/s11665-014-0958-z
[5] T. D. Ngo, A. Kashani, G. Imbalzano, and D. Hui, "Additive manufacturing (3D printing): a review of materials, methods, applications, and challenges," *Composites, Part B*, vol. 143, 2018, pp. 172–196, https://doi.org/10.1016/j.compositesb.2018.02.012
[6] H. Bikas, P. Stavropoulos, and G. Chryssolouris, "Additive manufacturing methods and modelling approaches: a critical review," *Int. J. Adv. Manuf. Technol.*, vol. 83, 2016, pp. 389–405, https://doi.org/10.1007/s00170-015-7576-2
[7] E. MacDonald and R. Wicker, "Multiprocess 3D printing for increasing component functionality," *Science*, vol. 353, 2016, pp. aaf2093, https://doi.org/10.1126/science.aaf2093
[8] S. Kannan and D. Senthilkumaran, "Investigating the influence of electroplating layer thickness on the tensile strength for fused deposition processed ABS thermoplastics," *Int. J. Eng. Technol.*, vol. 6, 2014, pp. 1047–1053.
[9] J. R. Jian, T. Kim, J. S. Park, J. Wang, and W. S. Kim, "High performance 3D printed electronics using electroless plated copper," *AIP Adv.*, vol. 7, 2017, pp. 035314, https://doi.org/10.1063/1.4979173
[10] T. K. Nguyen and B.-K. Lee, "Post-processing of FDM parts to improve surface and thermal properties," *Rapid Prototyping J.*, vol. 24, 2018, pp. 1091–1100, https://doi.org/10.1108/RPJ-12-2016-0207
[11] X. Su, X. Li, C. Y. A. Ong, T. S. Herng, Y. Wang, E. Peng, and J. Ding, "Metallization of 3D printed polymers and their application as a fully functional water-splitting system," *Adv. Sci.*, vol. 6, 2019, pp. 1801670, https://doi.org/10.1002/advs.201801670
[12] R. Bernasconi, G. Natale, M. Levi, M. Tironi and L. Magagnin, "Electroless plating of PLA and PETG for 3D printed flexible substrates," *ECS Trans.*, vol. 66, 2015, pp. 23–35, https://doi.org/10.1149/06619.0023ecst
[13] C. Kerr, D. Barker, and F. Walsh, "Electrolytic deposition (electroplating) of metals," *Trans. IMF*, vol. 80, 2002, pp. 67–72, doi:10.1080/00202967.2002.11871436

[14] S. Singamneni and O. Diegel, "Some recent developments and experiences with rapid manufacturing by indirect means," *AIJSTPME*, vol. 3, 2010, pp. 7–14.

[15] D. Weng and U. Landau, "Direct electroplating on nonconductors," *J. Electrochem. Soc.*, vol. 142, 1995, pp. 2598–2604, https://doi.org/10.1149/1.2050060

[16] P. F. Flowers, C. Reyes, S. Ye, M. J. Kim, and B. J. Wiley, "3D printing electronic components and circuits with conductive thermoplastic filament," *Addit. Manuf.*, vol. 18, 2017, pp. 156–163, https://doi.org/10.1016/j.addma.2017.10.002

[17] A. Islam, H. N. Hansen, P. T. Tang, "Direct electroplating of plastic for advanced electrical applications," *CIRP Ann.*, vol. 66, 2017, pp. 209–212, https://doi.org/10.1016/j.cirp.2017.04.124

[18] K. Angel, H. H. Tsang, S. S. Bedair, G. L. Smith and N. Lazarus, "Selective electroplating of 3D printed parts," *Addit Manuf.*, vol. 20, 2018, pp. 164–172, https://doi.org/10.1016/j.addma.2018.01.006

[19] D. Filonov, S. Kolen, A. Shmidt, Y. Shacham-Diamand, A. Boag, and P. Ginzburg, "Volumetric 3D-printed antennas, manufactured via selective polymer metallization," *Phys. Status Solidi.*, vol. 13, 2019, pp. 1800668, https://doi.org/10.1002/pssr.201800668

[20] N. Lazarus, S. S. Bedair, S. H. Hawasli, M. J. Kim, B. J. Wiley, and G. L. Smith, "Selective electroplating for 3D-printed electronics," *Adv. Mater. Technol.*, vol. 4, 2019, pp. 1900126, https://doi.org/10.1002/admt.201900126

[21] M. J. Kim, M. A. Cruz, S. Ye, A. L. Gray, G. L. Smith, N. Lazarus, C. J. Walker, H. H. Sigmarsson, and B. J. Wiley, "One-step electrodeposition of copper on conductive 3D printed objects," *Addit. Manuf.*, vol. 27, 2019, pp. 318–326, https://doi.org/10.1016/j.addma.2019.03.016

[22] S. Hawasli, H. Tsang, N. Lazarus, G. Smith, and E. Forsythe, "Improving conductivity of 3D printed conductive pastes for RF & high performance electronics," *Proc. IMWS-AMP 2018*, 2018, https://doi.org/10.1109/IMWS-AMP.2018.8457162

[23] J. J. Kelly, C. Tian, and A. C. West, "Leveling and microstructural effects of additives for copper electrodeposition," *J. Electrochem. Soc.*, vol. 146, 1999, pp. 2540–2545, https://doi.org/10.1149/1.1391968

[24] A. Agrawl and K. K. Sahu, "An overview of the recovery of acid from spent acidic solutions from steel and electroplating industries," *J. Hazard. Mater.*, vol. 171, 2009, pp. 61–75, https://doi.org/10.1016/j.jhazmat.2009.06.099

[25] Z. Chen and S. Liu, "Simulation of copper electroplating fill process of through silicon via," Proc. ICEPT 2010, 2010, pp. 433–437, https://doi.org/10.1109/ICEPT.2010.5583828

[26] C. A. Deckert, "Electroless copper plating a review: part I," *Plat. Surf. Finish.*, vol. 171, 1995, pp. 48–55.

[27] S. Ghosh, "Electroless copper deposition: a critical review," *Thin Solid Films*, vol. 669, 2019, pp. 641–658, https://doi.org/10.1016/j.tsf.2018.11.016

[28] C. R. Shipley, "Historical highlights of electroless plating," *Plat. Surf. Finish.*, vol. 71, 1984, pp. 24–27.

[29] F. Hanna, Z. A. Hamid, and A. A. Aal, "Controlling factors affecting the stability and rate of electroless copper plating," *Mat. Lett.*, vol. 58, 2003, pp. 104–109, https://doi.org/10.1016/S0167-577X(03)00424-5

[30] S. Siau, J. De Baets, A. Van Calster, L. Heremans, and S. Tanghe, "Processing quality results for electroless/electroplating of high-aspect ratio plated through holes in industrially produced printed circuit boards," *Microelectron. Reliab.*, vol. 45, 2005, pp. 675–687, https://doi.org/10.1016/j.microrel.2004.07.009

[31] G. P. Le Sage, "3D printed waveguide slot array antennas," *IEEE Access*, vol. 4, 2016, pp. 1258–1265, https://doi.org/10.1109/ACCESS.2016.2544278

[32] S. Olivera, H. B. Muralidhara, K. Venkatesh, K. Gopalakrishna, and C. S. Vivek, "Plating on acrylonitrile-butadiene-styrene (ABS) plastic: a review," *J. Mater. Sci.*, vol. 51, 2016, pp. 3657–3674, https://doi.org/10.1007/s10853-015-9668-7

[33] R. Bernasconi, E. Carrara, M. Hoop, F. Mushtaq, X. Chen, B. J. Nelson, S. Pane, C. Credi, M. Levi, and L. Magagnin, "Magnetically navigable 3D printed multifunctional microdevices for environmental applications," *Addit. Manuf.*, vol. 28, 2019, pp. 127–135, https://doi.org/10.1016/j.addma.2019.04.022

[34] X. Wang, Q. Guo, X. Cai, S. Zhou, B. Kobe, and J. Yang, "Initiator-integrated 3D printing enables the formation of complex metallic architectures," *ACS Appl. Mater. Interfaces*, vol. 6, 2014, pp. 2583–2587, https://doi.org/10.1021/am4050822

[35] A. Equbal and A. K. Sood, "Investigations on metallization in FDM build ABS part using electroless deposition method," *J. Manuf. Proc.*, vol. 19, 2015, pp. 22–31, https://doi.org/10.1016/j.jmapro.2015.03.002

[36] Y. Li, C. Wang, H. Yuan, N. Liu, H. Zhao, and X. Li, "A 5G MIMO antenna manufactured by 3-D printing method," *IEEE Antennas Wireless Propag. Lett.*, vol. 16, 2017, pp. 657–660, https://doi.org/10.1109/LAWP.2016.2596297

[37] F. Formanek, N. Takeyasu, T. Tanaka, K. Chiyoda, A. Ishikawa, and S. Kawata, "Selective electroless plating to fabricate complex three dimensional metallic micro/nanostructures," *Appl. Phys. Lett.*, vol. 88, 2006, pp. 083110, https://doi.org/10.1063/1.2178261

[38] J. Li, Y. Wang, G. Xiang, H. Liu, and J. He, "Hybrid additive manufacturing method for selective plating of freeform circuitry on 3D printed plastic structure," *Adv. Mater. Technol.*, vol. 4, 2019, pp. 1800529, https://doi.org/10.1002/admt.201800529

[39] A. Ryspayeva, T. D. A. Jones, S. R. Khan, M. N. Esfahani, M. P. Shuttleworth, R. A. Harris, R. W. Kay, M. P. Y. Desmulliez, and J. Marques-Hueso, "Selective metallization of 3D printable thermoplastic polyurethanes," *IEEE Access*, vol. 7, 2019, pp. 104947–104955, https://doi.org/10.1109/ACCESS.2019.2931594

[40] C. Gath and D. Drummer, "Circuit board application to additive manufacturing components by laser-direct-structuring," *Proc. 2016 MID*, 2016, https://doi.org/10.1109/ICMID.2016.7738926

[41] J. Zhan, T. Tamura, X. Li, Z. Ma, M. Sone, M. Yoshino, S. Umezu, and H. Sato, "Metal-plastic hybrid 3D printing using catalyst-loaded filament and electroless plating," *Addit. Manuf.*, vol. 36, 2020, pp. 101556, https://doi.org/10.1016/j.addma.2020.101556

[42] P. BalaRamesh, P. Venkatesh, and A. Sampath, "The effects of 'green' complexing agents on electroless deposition of copper from methanesulfonate baths," *Surf. Coat. Technol.*, vol. 276, 2015, pp. 233–238, https://doi.org/10.1016/j.surfcoat.2015.06.066

[43] A. K. Au, W. Huynh, L. F. Horowitz, and A. Folch, "3D-printed microfluidics," *Angew. Chem. Int. Ed.*, vol. 55, 2016, pp. 3862–3881, https://doi.org/10.1002/anie.201504382

[44] S. Wu, C. Yang, W. Hsu, and L. Lin, "3D-printed microelectronics for integrated circuitry and passive wireless sensors," *Microsyst. Nanoeng.*, vol. 1, 2015, pp. 15013, https://doi.org/10.1038/micronano.2015.13

[45] M. D. Dickey, R. C. Chiechi, R. J. Larsen, E. A. Weiss, D. A. Weitz, and G. M. Whitesides, "Eutectic gallium-indium (EGaIn): a liquid metal alloy for the formation of stable structures in microchannels at room temperature," *Adv. Funct. Mater.*, vol. 18, 2008, pp. 1097–1104, https://doi.org/10.1002/adfm.200701216

[46] Y. Yu, J. Lu, and J. Liu, "3D printing for functional electronics by injection and package of liquid metals into channels of mechanical structures," *Mater. Des.*, vol. 122, 2017, pp. 80–89, https://doi.org/10.1016/j.matdes.2017.03.005

[47] S. Cheng and Z. Wu, "Microfluidic electronics," *Lab Chip*, vol. 12, 2012, pp. 2782–2791, https://doi.org/10.1039/c2lc21176a

[48] M. A. H. Khondoker and D. Sameoto, "Fabrication methods and applications of microstructured gallium based liquid metal alloys," *Smart Mater. Struct.*, vol. 25, 2016, pp. 093001, https://doi.org/10.1088/0964-1726/25/9/093001

[49] C. Ladd, J. So, J. Muth, and M. D. Dickey, "3D printing of free standing liquid metal microstructures," *Adv. Mater.*, vol. 25, 2013, pp. 5081–5085, https://doi.org/10.1002/adma.201301400

[50] S. Park, K. Mondal, R. M. Treadway, V. Kumar, S. Ma, J. D. Holbery, and M. D. Dickey, "Silicones for stretchable and durable soft devices: beyond Sylgard 184," *ACS Appl. Mater. Interfaces*, vol. 10, 2018, pp. 11261–11268, https://doi.org/10.1021/acsami.7b18394

[51] Q. Chen, J. Zhao, J. Ren, L. Rong, P. Cao, and R. C. Advincula, "3D printed multifunctional, hyperelastic silicone rubber foam," *Adv. Funct. Mater.*, vol. 29, 2019 pp. 1900469, https://doi.org/10.1002/adfm.201900469

[52] F. Liravi and E. Toyserkani, "Additive manufacturing of silicone structures: a review and prospective," *Addit. Manuf.*, vol. 24, 2018, pp. 232–242, https://doi.org/10.1016/j.addma.2018.10.002

[53] A. Fassler and C. Majidi, "Soft-matter capacitors and inductors for hyperelastic strain sensing and stretchable electronics," *Smart Mater. Struct.*, vol. 22, 2013, pp. 055023, https://doi.org/10.1088/0964-1726/22/5/055023

[54] N. Lazarus, C. D. Meyer, S. S. Bedair, H. Nochetto, and I. M. Kierzewski, "Multilayer liquid metal stretchable inductors," *Smart Mater. Struct.*, vol. 23, 2014, pp. 085036, https://doi.org/10.1088/0964-1726/23/8/085036

[55] N. Lazarus, S. S. Bedair, and G. L. Smith, "Creating 3D printed magnetic devices with ferrofluids and liquid metals," *Addit. Manuf.*, vol. 26, 2019, pp. 15–21, https://doi.org/10.1016/j.addma.2018.12.012

[56] V. Bharambe, D. P. Parekh, C. Ladd, K. Moussa, M. D. Dickey, and J. J. Adams, "Vacuum-filling of liquid metals for 3D printed RF antennas," *Addit. Manuf.*, vol. 18, 2017, pp. 221–227, https://doi.org/10.1016/j.addma.2017.10.012

[57] C. Y. Lai, C. F. Cheong, J. S. Mandeep, H. B. Abdullah, N. Amin, and K. W. Lai, "Synthesis and characterization of silver nanoparticles and silver inks: Review on the past and recent technology roadmaps," *JMEPEG*, vol. 23, 2004, pp. 3541–3550, https://doi.org/10.1007/s11665-014-1166-6

[58] P. Jiang, Z. Ji, X. Zhang, Z. Liu, and X. Wang, "Recent advances in direct ink writing of electronic components and functional devices," *Prog. Addit. Manuf.*, vol. 3, 2018, pp. 65–86, https://doi.org/10.1007/s40964-017-0035-x

[59] C. Goth, S. Putzo, and J. Franke, "Aerosol jet printing on rapid prototyping materials for fine pitch electronic applications," *Proc. ECTC 2011*, 2011, pp. 1211–1216, https://doi.org/10.1109/ECTC.2011.5898664

[60] J. Perelaer, R. Jani, M. Grouchko, A. Kamyshny, S. Magdassi, and U. S. Schubert, "Plasma and microwave flash sintering of a tailored silver nanoparticle ink, yielding 60% bulk conductivity on cost-effective polymer foils," *Adv. Mater.*, vol. 24, 2012, pp. 3993–3998, https://doi.org/10.1002/adma.201200899

[61] M. Braginsky, V. Tikare, and E. Olevsky, "Numerical simulation of solid state sintering," *Int. J. Solids Struct.*, vol. 42, 2005, pp. 621–636, https://doi.org/10.1016/j.ijsolstr.2004.06.022

[62] A. Roshanghias, M. Krivec, and M. Baumgart, "Sintering strategies for inkjet printed metallic traces in 3D printed electronics," *Flex. Print. Electron.*, vol. 2, 2017, pp. 045002, https://doi.org/10.1088/2058-8585/aa8ed8

[63] Y. Liu, M. Tsai, Y. Pai, and W. Hwang, "Control of droplet formation by operating waveform for inks with various viscosities in piezoelectric inkjet printing," *Appl. Phys. A*, vol. 111, 2013, pp. 509–516, https://doi.org/10.1007/s00339-013-7569-7

[64] I. H. Choi, J. Kim, "A pneumatically driven inkjet printing system for highly viscous microdroplet formation," *Micro and Nano Syst. Lett.*, vol. 4, 2016, pp. 1–7, https://doi.org/10.1186/s40486-016-0030-x

[65] E. Jabari and E. Toyserkani, "Micro-scale aerosol-jet printing of graphene interconnects," *Carbon*, vol. 91, 2015, pp. 321–239, https://doi.org/10.1016/j.carbon.2015.04.094

[66] T. Seifert, E. Sowade, F. Roscher, M. Wiemer, T. Gessner, and R. R. Baumann, "Additive manufacturing technologies compared: morphology of deposits of silver ink using inkjet and aerosol jet printing," *Ind. Eng. Chem. Res.*, vol. 54, 2015, pp. 769–779, https://doi.org/10.1021/ie503636c

[67] L. Friedrich and M. Begley, "In situ characterization of low-viscosity direct ink writing: stability, wetting and rotational flows," *J. Colloid Interface Sci.*, vol. 529, 2018, pp. 599–609, https://doi.org/10.1016/j.jcis.2018.05.110

[68] A. Shen, D. Caldwell, A. W. K. Ma, and S. Dardona, "Direct write fabrication of high-density parallel silver interconnects," *Addit. Manuf.*, vol. 22, 2018, pp. 343–350, https://doi.org/10.1016/j.addma.2018.05.010

[69] Y. Mou, Y. Zhang, H. Cheng, Y. Peng and M. Chen, "Fabrication of highly conductive and flexible printed electronics by low temperature sintering reactive silver ink," *Appl. Surf. Sci.*, vol. 459, 2018, pp. 249–256, https://doi.org/10.1016/j.apsusc.2018.07.187

[70] S. B. Walker and J. A. Lewis, "Reactive silver inks for patterning high-conductivity features at mild temperatures," *J. Am. Chem. Soc.*, vol. 134, 2013, pp. 1419–1421, https://doi.org/10.1021/ja209267c

[71] Y. Akiyama, T. Sugiyama, and H. Kawasaki, "Contribution of ligand oxidation products to high durability of copper films prepared from low-sintering-temperature copper ink on polymer substrates," *Adv. Eng. Mater.*, vol. 19, 2018, pp. 1700259, https://doi.org/10.1002/adem.201700259

[72] W. Gao, H. Gong, J. He, A. Thomas, L. Chan, and S. Li, "Oxidation behavior of Cu thin films on Si wafer at 175–400°C," *Mater. Lett.*, vol. 51, 2001, pp. 78–84, https://doi.org/10.1016/S0167-577X(01)00268-3

[73] N. Lazarus, C. D. Meyer, S. S. Bedair, X. Song, L. M. Boteler, and I. M. Kierzewski, "Thick film oxidation of copper in an electroplated MEMS process," *J. Micromech. Microeng.*, vol. 23, 2013, pp. 065017, https://doi.org/10.1088/0960-1317/23/6/065017

[74] Y. Kwon, Y. Lee, S. Kim, K. Lee, and Y. Choa, "Full densification of inkjet-printed copper conductive tracks on a flexible substrate utilizing a hydrogen plasma sintering," *Appl. Surf. Sci.*, vol. 396, 2017, pp. 1239–1244, https://doi.org/10.1016/j.apsusc.2016.11.122

[75] R. Dharmadasa, M. Jha, D. A. Amos, and T. Druffel, "Room temperature synthesis of a copper ink for the intense pulsed light sintering of conductive copper films," *ACS Appl. Mater. Interfaces*, vol. 5, 2013, pp. 12337–13234, https://doi.org/10.1021/am404226e

[76] K. Chou, K. Huang, and H. Lee, "Fabrication and sintering effect on the morphologies and conductivity of nano-Ag particle films by the spin coating method," *Nanotechnology*, vol. 16, 2005, pp. 779–784, https://doi.org/10.1088/0957-4484/16/6/027

[77] H. Kim, S. R. Dhage, D. Shim, and H. T. Hahn, "Intense pulsed light sintering of copper nanoink for printed electronics," *Appl. Phys. A*, vol. 97, 2009, pp. 791–798, https://doi.org/10.1007/s00339-009-5360-6

[78] K. Maekawa, K. Yamasaki, T. Niizeki, M. Mita, Y. Matsuba, N. Terada, and H. Saito, "High-speed laser plating on Cu leadframe using Ag nanoparticles," *Proc. ECTC 2010*, 2010, https://doi.org/10.1109/ECTC.2010.5490933

[79] M. Hummelgard, R. Zhang, H. Nilsson, and H. Olin, "Electrical sintering of silver nanoparticle ink studied by in-situ TEM probing," *PLoS One*, vol. 6, 2011, pp. e17209, https://doi.org/10.1371/journal.pone.0017209

[80] Q. Mu, M. Lei, D. J. Roach, C. K. Dunn, X. Kuang, C. Yuan, T. Wang, and H. J. Qi, "Intense pulsed light sintering of thick conductive wires on elastomeric dark substrate for hybrid 3D printing applications," *Smart Mater. Struct.*, vol. 27, 2018, pp. 115007, https://doi.org/10.1088/1361-665X/aae11f

[81] J. Gonzalez-Gutierrez, S. Cano, S. Schuschnigg, C. Kukla, J. Sapkota, and C. Holzer, "Additive manufacturing of metallic and ceramic components by the material extrusion of highly-filled polymers: a review and future perspectives," *Mater.*, vol. 11, 2018, pp. 840, https://doi.org/10.3390/ma11050840

[82] J. A. Cardenas, H. Tsang, H. Tong, H. Abuzaid, K. Price, M. A. Cruz, B. J. Wiley, A. D. Franklin, and N. Lazarus, "Flash ablation metallization of conductive thermoplastics," *Addit. Manuf.*, vol. 36, 2020, pp. 101409, https://doi.org/10.1016/j.addma.2020.101409

[83] Z. Zhao, A. Mamidanna, C. Lefky, O. Hildreth, and T. L. Alford, "A percolative approach to investigate electromigration failure in printed Ag structures," *J. Appl. Phys.*, vol. 120, 2016, pp. 125104, https://doi.org/10.1063/1.4963755

[84] W. Lin and F. Ouyang, "Electromigration behavior of screen-printing silver nanoparticles interconnects," *JOM*, vol. 71, 2019, pp. 3084–3093, https://doi.org/10.1007/s11837-019-03627-0

[85] Z. J. Radzimski, "Directional copper deposition using dc magnetron self-sputtering," *J. Vac. Sci. Technol. B*, vol. 16, 1998, pp. 1102–1106, https://doi.org/10.1116/1.590016

[86] A. Baptista, F. Silva, J. Porteiro, J. Miguez, and G. Pinto, "Sputtering physical vapour deposition (PVD) coatings: a critical review on process improvement and market trend demands," *Coatings*, vol. 8, 2018, pp. 402, https://doi.org/10.3390/coatings8110402

[87] A. H. Simon, "Sputter processing," in K. Seshan and D. Schepis (Eds.), *Handbook of Thin Film Deposition Fourth Edition*, Elsevier, Amsterdam, 2018, pp. 195–230.

[88] S. D. Senturia, *Microsystem Design*, Kluwer Academic Publishing, New York, 2001.

[89] D. Depla, S. Mahieu, and J. E. Greene, "Sputter Deposition Processes," in P. M. Martin (Ed), *Handbook of Deposition Technologies for Films and Coatings 3rd* Edition, William Andrew, Boston, 2010, pp. 253–296.

[90] M. Huff, "MEMS Fabrication," *Sens. Rev.*, vol. 22, 2002, pp. 18–33, https://doi.org/10.1108/02602280210697087

[91] S. M. Rossnagel, "Directional and ionized physical vapor deposition for microelectronics applications," *J. Vac. Sci. Technol. B*, vol. 16, 1998, pp. 2585–2608, https://doi.org/10.1116/1.590242

[92] L. Abelmann and C. Lodder, "Oblique evaporation and surface diffusion," *Thin Solid Films*, vol. 305, 1997, pp. 1–21, https://doi.org/10.1016/S0040-6090(97)00095-3

[93] M. I. M. Ghazali, E. Gutierrez, J. C. Meyers, A. Kaur, B. Wright, and P. Chahal, "Affordable 3D printed microwave antennas," *Proc. ECTC 2015*, vol. 2015, 2015, pp. 240–246, https://doi.org/10.1109/ECTC.2015.7159599

[94] M. I. M. Ghazali, S. Karuppuswami, A. Kaur, and P. Chahal, "3-D printed air substrates for the design and fabrication of RF components," *IEEE Trans. on Comp., Packag., Manuf. Technol.*, vol. 7, 2017, pp. 982–989, https://doi.org/10.1109/TCPMT.2017.2686706

[95] C. R. Garcia, R. C. Rumpf, H. H. Tsang, and J. H. Barton, "Effects of extreme surface roughness on 3D printed horn antenna," *Electron. Lett.*, vol. 49, 2013, pp. 734–736, https://doi.org/10.1049/el.2013.1528

[96] A. Afshar and D. Mihut, "Enhancing durability of 3D printed polymer structures by metallization," *J. Mater. Sci. Technol.*, vol. 53, 2020, pp. 185–191, https://doi.org/10.1016/j.jmst.2020.01.072

[97] W. M. Abbott, C. P. Murray, S. N. Lochlainn, F. Bello, C. Zhong, C. Smith, E. K. McCarthy, C. Downing, D. Daly, A. K. Petford-Long, C. McGuinness, I. I. Chunin, J. F. Donegan, and D. McCloskey, "Comparison of metal adhesion layers for Au films in thermoplasmonic applications," *ACS Appl. Mater. Interfaces*, vol. 12, 2020, pp. 13503–13509, https://doi.org/10.1021/acsami.9b22279

[98] C. W. Tan and J. Miao, "Optimization of sputtered Cr/Au thin films for diaphragm-based MEMS applications," *Thin Solid Films*, vol. 517, 2009, pp. 4921–4925, https://doi.org/10.1016/j.tsf.2009.03.151

[99] J. White, C. Tenore, A. Pavich, R. Scherzer, and S. Stagon, "Environmentally benign metallization of material extrusion technology 3D printed acrylonitrile butadiene styrene parts using physical vapor deposition," *Addit. Manuf.*, vol. 22, 2018, pp. 279–285, https://doi.org/10.1016/j.addma.2018.05.016

[100] K. V. Hoel, S. Kristoffersen, J. Moen, K. G. Kjelgard, and T. S. Lande, "Broadband antenna design using different 3D printing technologies and metallization processes," *Proc. EuCAP 2016*, 2016, https://doi.org/10.1109/EuCAP.2016.7481620

[101] H. W. Tan, T. Tran, and C. K. Chua, "A review of printed passive electronic components through fully additive manufacturing methods," *Virtual Phys. Prototyping*, vol. 11, 2016, pp. 271–277, https://doi.org/10.1080/17452759.2016.1217586

[102] W. Liang, L. Raymond, and J. Rivas, "3-D-printed air-core inductors for high-frequency power converters," *IEEE Trans. Power Electron.*, vol. 31, 2016, pp. 52–64, https://doi.org/10.1109/TPEL.2015.2441005

[103] W. Liang, L. Raymond, M. Praglin, D. Biggs, F. RIghetti, M. Cappelli, B. Holman, and J. R. Davila, "Low-mass RF power inverter for CubeSat applications using 3-D printed inductors," *IEEE J. Emerging Sel. Top. Power Electron.*, vol. 5, 2017, pp. 880–890, https://doi.org/10.1109/JESTPE.2016.2644644

[104] T. Hou, J. Xu, W. S. Elkhuizen, C. C. L. Wang, J. Jiang, J. M. P. Geraedts, and Y. Song, "Design of 3D wireless power transfer system based on 3D printed electronics," *IEEE Access*, vol. 7, 2019, pp. 94793–94805, https://doi.org/10.1109/ACCESS.2019.2928948

[105] L. Li, R. Abedini-Nassab, and B. B. Yellen, "Monolithically integrated Helmholtz coils by 3-dimensional printing," *Appl. Phys. Lett*, vol. 104, 2014, pp. 253–505, https://doi.org/10.1063/1.4885441

[106] B. K. Bose, "Power electronics – a technology review," *Proc. IEEE*, vol. 80, 1992, pp. 1303–1334, https://doi.org/10.1109/5.158603

[107] S. Y. R. Hui, W. Zhong, and C. K. Lee, "A critical review of recent progress in mid-range wireless power transfer," *IEEE Trans. Power Electron.*, vol. 29, 2014, pp. 4500–4511, https://doi.org/10.1109/TPEL.2013.2249670

[108] E. Aguilera, J. Ramos, D. Espalin, F. Cedillos, D. Muse, R. Wicker and E. MacDonald, "3D printing of electro mechanical systems," *Proc. 24th Int. Solid Freeform Fabrication Symp.*, 2013, pp. 950–961.

[109] K. Park, Z. Jiang, and S. Hwang, "Design and analysis of a novel microspeaker with enhanced low-frequency SPL and size reduction," *Appl. Sci.*, vol. 10, 2020, pp. 8902, https://doi.org/10.3390/app10248902

[110] P. Constantinou and S. Roy, "A 3D printed electromagnetic nonlinear vibration energy harvester," *Smart Mater. Struct.*, vol. 25, 2016, pp. 095053, https://doi.org/10.1088/0964-1726/25/9/095053

[111] D. W. Lee, K. Hwang, and S. X. Wang, "Fabrication and analysis of high-performance integrated solenoid inductor with magnetic core," *IEEE Trans. Magn.*, vol. 44, 2008, pp. 4089–4095, https://doi.org/10.1109/TMAG.2008.2003398

[112] O. F. Hikmat and M. S. M. Ali, "RF MEMS inductors and their applications – A review," *J. Microelectromech. Syst.*, vol. 26, 2017, pp. 17–44, https://doi.org/10.1109/JMEMS.2016.2627039

[113] J. A. Rogers, T. Someya, and Y. Huang, "Materials and mechanics for stretchable electronics," *Science*, vol. 327, 2010, pp. 1603–1607, https://doi.org/10.1126/science.1182383

[114] J. W. Boley, E. L. White, G. T.-C. Chiu, and R. K. Kramer, "Direct writing of gallium-indium alloy for stretchable electronics," *Adv. Funct. Mater.*, vol. 24, 2014, pp. 3501–3507, https://doi.org/10.1002/adfm.201303220

[115] D. Kim, N. Lu, R. Ma, Y. Kim, R. Kim, S. Wang, J. Wu, S. M. Won, H. Tao, A. Islam, K. J. Yu, T. Kim, R. Chowdhury, M. Ying, L. Xu, M. Li, H. Chung, H. Keum, M. McCormick, P. Liu, Y. Zhang, F. G. Omenetto, Y. Huang, T. Coleman, and J. A. Rogers, "Epidermal electronics," *Science*, vol. 333, 2011, pp. 838–843, https://doi.org/10.1126/science.1206157

[116] R. Kim, H. Tao, T. Kim, Y. Zhang, S. Kim, B. Panilaitis, M. Yang, D. Kim, Y. H. Jung, B. H. Kim, Y. Li, Y. Huang, F. G. Omenetto, and J. A. Rogers, "Materials and designs for wirelessly powered implantable light-emitting systems," *Small*, vol. 8, 2012, pp. 2812–2818, https://doi.org/10.1002/smll.201200943

[117] S. H. Jeong, K. Hjort, and Z. Wu, "Tape transfer atomization patterning of liquid alloys for microfluidic stretchable wireless power transfer," *Sci. Rep.*, vol. 5, 2015, pp. 8419, https://doi.org/10.1038/srep08419

[118] W. Su, S. A. Nauroze, B. Ryan and M. M. Tentzeris, "Novel 3D printed liq-uid-metal-alloy microfluidics-based zigzag and helical antennas for origami reconfigu-rable antenna 'trees'", *2017 IEEE MTT-S International Microwave Symposium (IMS)*, 2017, pp. 1579–1582, https://doi.org/10.1109/MWSYM.2017.8058933

[119] E. Siman-Tov, V. F. Tseng, S. S. Bedair, and N. Lazarus, "Increasing the range of wire-less power transmission to stretchable electronics," *IEEE Trans. Microwave Theory Tech.*, vol. 66, 2018, pp. 5021–5030, https://doi.org/10.1109/TMTT.2018.2859948

[120] Y. Yan, J. Moss, K. D. T. Ngo, Y. Mei, and G. Lu, "Additive manufacturing of toroid inductor for power electronics applications," *IEEE Trans. Ind. Appl.*, vol. 53, 2017, pp. 5709–5714, https://doi.org/10.1109/TIA.2017.2729504

[121] C. Shemelya, F. Cedillos, E. Aguilera, D. Espalin, D. Muse, R. Wicker, and E. MacDonald, "Encapsulated copper wire and copper mesh capacitive sensing for 3-D printing applications," *IEEE Sens. J.*, vol. 15, 2015, pp. 1280–1286, https://doi.org/10.1109/JSEN.2014.2356973

[122] K. M. M. Billah, J. L. Coronel, M. C. Halbig, R. B. Wicker, and D. Espalin, "Electrical and thermal characterization of 3D printed thermoplastic parts with embedded wires for high current-carrying applications," *IEEE Access*, vol. 7, 2019, pp. 18799–18810, https://doi.org/10.1109/ACCESS.2019.2895620

[123] R. Matsuzaki, T. Kanatani, and A. Todoroki, "Multi-material additive manufacturing of polymers and metals using fused filament fabrication and electroforming," *Addit. Manuf.*, vol. 29, 2019, pp. 100812, https://doi.org/10.1016/j.addma.2019.100812

6 Printing Electrically Conductive Patterns on Polymeric and 3D-Printed Systems

Lorenzo Migliorini, Tommaso Santaniello, and Paolo Milani
Centro Interdisciplinare Materiali e
Interfacce Nanostrutturati (CIMAINA),
Università degli Studi di Milano, Italy

6.1 INTRODUCTION

While traditional electronics and robotics rely on the use of rigid materials and flat platforms, rapidly developing technological needs ask for soft devices based on stretchable, flexible and deformable materials, able to adapt their size and shape according to different stimuli and duties as well as to withstand shocks [1–3]. This demand arises from the necessity to safely and efficiently interact with soft biological tissues and materials (like the human skin or the leaves of a plant) and to operate in harsh, extreme, or unknown environments [4–7]. Soft sensors and actuators, energy harvesters, as well as soft electronics platforms can play a key role in many forefront applications, such as real-time diagnostic, minimally invasive surgery, environmental monitoring and recognition, and precision agriculture [8–11]. Both synthetic and natural-derived materials can be of interest: the former usually characterized by better mechanical properties, while the latter benignly degrading in the environment [12,13].

To develop these functional devices and systems, different kinds of electronic elements have to be integrated onto soft substrates. The fabrication of conductive, resilient, well-adherent and durable metal electrodes coupled with soft polymeric materials represents a challenging technological task in order to achieve the realization of soft electronics devices. The mechanical mismatch between the conductive metallic layers and the polymeric substrates usually leads to low adhesion and delamination under deformation [14]. Moreover, with the advent of 3D printing and additive manufacturing, complex-shape and uncommon objects can be fabricated with different kinds of materials [15–18], however they are usually passive elements lacking functional properties. A high-resolution metallization patterning of these substrates results to be critical because of their three-dimensional full-and-empty structure; the integration of electronics on these objects can potentially lead to complex architectures characterized by functional 3D shapes improved with sensing energy management and communication capabilities.

Supersonic Cluster Beam Deposition (SCBD) is an additive technology based on the production of intense and highly collimated nanoparticle beams that enables the large-scale and high-throughput integration of nanostructured functional materials on a wide variety of substrates, including microfabricated platforms, smart nanocomposites and fragile materials [19–25]. A picture and a schematic of the SCBD apparatus is reported in Figure 6.1. A supersonically accelerated beam of neutral metal clusters is aerodynamically collimated and directed onto a substrate of

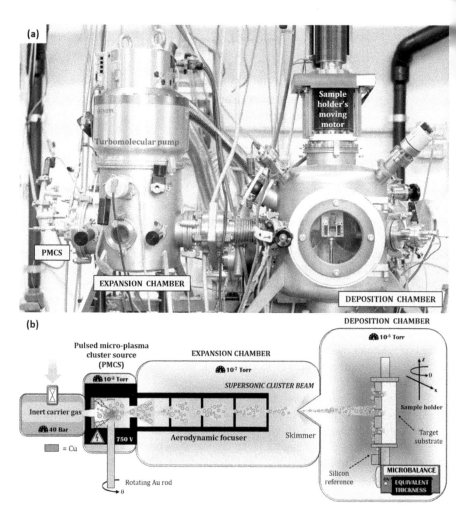

FIGURE 6.1   (a) Photograph and (b) scheme of the Supersonic Cluster Beam Deposition apparatus. It can be seen how the metallic clusters form in the pulsed micro-plasma cluster source and are delivered throughout the expansion and the deposition chambers, finally reaching the target's surface (from Ref. [33]).

choice to generate thin cluster-assembled metal layers deposited on its surface. The nanoparticles are generated in situ in a pulsed micro-plasma cluster source (PMCS) from the sputtering of a solid precursor. The sputtered atoms condense in clusters that are carried by an inert gas throughout an aerodynamic focuser (size distribution from 3 to 10 nm). The resulting gas beam, seeded with the clusters, undergoes a supersonic expansion and it finally reaches the deposition chamber, where the cluster beam intercepts the substrates, mounted on a sample holder that can be moved and rotated in order to allow the metallization of large areas and three-dimensional objects. The low kinetic energy of the clusters (0.5 eV per atom) allows them to deposit or implant on the surface of soft polymeric materials without inducing any chemical or physical alteration [26–28]. SCBD is particularly suited for metallization of soft polymeric materials and 3D printed objects. In fact, it allows the deposition of thin conductive films with a cluster-assembled structure that renders them resilient to deformation and characterized by a robust adhesion with the substrate. Micro-patterning of different kinds of flat films and materials can be easily achieved by the use of stencil masks [26]. The additive manufacturing nature of this technique allows the patterning of three-dimensional objects as well, thanks to the possibility to free motion and rotation of the substrate during the deposition process [29–31]. With respect to other metallization techniques, such as physical vapor deposition or magnetron sputtering, the film deposited with SCBD possess a cluster-assembled morphology characterized by a lower density, resulting in lightness and in the presence of many nanostructured inter-particle joints. Aerosol jet printing (AJP) is another technique widely used in the context of printed electronics and it employs conductive inks to deposit nanostructured paths on 2D and 3D surfaces (with a thickness range spanning between 100 nm and tens of μm) [32]. On the other side, SCBD has the great advantage of being a dry technique, removing the issues of the solvent's toxicity and disposal, as well as the NPs solubilization and stabilization. Moreover, the clusters can be implanted underneath the surface of a soft substrate, providing electrodes much more stable and resilient to deformation (as it is reported in the next sections). The wide deposition rate (40–700 Å/s) allows to precisely target small thickness values from 5 to hundreds of nm, consequently tuning the sheet resistance (from ca. $10^6\ \Omega$ to $10^1\ \Omega$).

In the next sections, some examples are shown where SCBD has been employed to deposit conductive traces made of cluster-assembled Au on different polymeric materials. Transparent sheets of polypropylene and cellulose acetate have been metallized to obtain flexible conductors or bending sensors (Section 6.2). Au clusters have been implanted on elastomeric materials such as polydimethylsiloxane (PDMS) to form soft conductive paths resilient under mechanical stretching (Section 6.3). Soft ionogels, characterized by ionic conductivity, have been metallized as well with SCBD technique and the resulting ionogel-metal nanocomposite materials can show a variety of interesting properties such as electro-mechanical actuation, energy storage and piezoelectricity (Section 6.4). Some examples are also shown where SCBD was combined with fused filament fabrication (FFF) to obtain an integrated additive manufacturing approach that allowed to obtain 3D-printed electronic devices integrating different sensing elements, resistors and electrodes (Section 6.5).

6.2   CONDUCTIVE TRACES ON TRANSPARENT FLEXIBLE SUBSTRATES

Flexible and transparent electronics are of fundamental importance for the development of efficient, stretchable and wearable optoelectronic devices such as organic photovoltaics (OPVs), organic light-emitting diodes (OLEDs), e-readers, touch screens and electronic papers [34–39]. Transparent conductive films (TCFs) should exhibit high optical transmittance and high electronic conduction even under severe deformation. Novel emerging materials such as conductive polymers, metal grids, carbon nanotubes (CNTs), graphene and random networks of metal nanowires and nanoparticles have been explored [40–43] in combination with printing techniques, providing a promising solution to the scale-up problem. The formulation of printable conductive inks is one of the main crucial tasks to be considered: the composition and the physical properties such as viscosity and surface tension are pivotal for achieving high printing accuracy and resolution. Besides the transparency, flexibility is also a highly desirable feature for conductive films. Flexible electrodes can retain their conductivity under deformation, and they can also be employed as strain/bending sensors: Soft flexible sensors can be attached onto the clothing and even directly mounted on the human skin or plants leaves for the real-time monitoring of physiological activities. Recently, several types of strain sensors have been proposed by using nanomaterials coupled with flexible and stretchable polymers [44–48].

SCBD can be employed to deposit Au conductive traces to obtain flexible and/or transparent electrodes able to work under mechanical deformation and to operate as bending sensors. As an example, 150-nm-thick Au films have been deposited on industrial polypropylene (PP), obtaining conductive paths with a sheet resistance of ca. 25 Ω and a linear I/V behavior up to ca. 1 V and 80 mA. Their resiliency under bending was checked up to values of 45° (Figure 6.2a). The employment of SCBD also allowed the development of TCFs and bending sensors, by depositing cluster-assembled Au electrodes on flexible, transparent and fully biodegradable cellulose acetate (CA) sheets [49]. The thickness and the conductance of the growing Au films were measured in situ during the deposition process (percolation curve in Figure 6.2b) and this allowed to obtain several samples characterized by different resistance values, ranging from 4 Ω to 1 kΩ. All of them showed linear I/V trends up to applied potentials of 2 V. The samples with the thinnest cluster-assembled Au layers (11 nm) possessed a high transmittance in the visible range of ca. 70%. Such value exponentially decreased for the samples with thicker Au layers (Figure 6.2c).

6.3   CONDUCTIVE TRACES IMPLANTED IN ELASTOMERIC/STRETCHABLE MATERIALS

Stretchable electrodes consisting of metallic paths on elastomeric substrates are of primary interest due to the extremely high mechanical deformation that this class of systems can undergo [50,51]. PDMS is an example of widely used elastomeric material that couples biocompatibility with mechanical properties and machinability suitable for rapid prototyping [52]. Many metallization techniques have been proposed and employed to deposit conductive electrodes on the surface of PDMS films, such as metal vapor deposition or ion metal implantation [53–55]; however, the use of

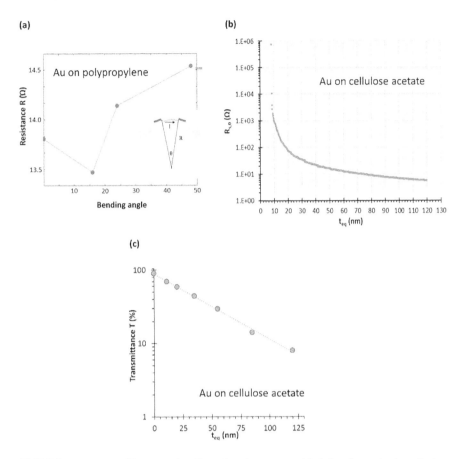

FIGURE 6.2 (a) Resiliency to bending of a cluster-assembled Au electrode deposited on polypropylene. (b) Percolation curve of a cluster-assembled Au film deposited on cellulose acetate. (c) Optical transmittance of the cellulose acetate sheets provided with Au electrodes with different thickness values (up to 125 nm).

these electrodes is usually hampered by delamination of the conducting layers. This is caused by the huge mechanical mismatch between the rigid metallic layers and the stretchable polymeric substrate (a Young's modulus of 50–100 GPa compared to 0.2–1 MPa), and this represents the primary issue to be addressed in order to obtain compliant and resilient electrodes on stretchable elastomeric materials [14].

SCBD can be employed to fabricate cluster-assembled Au conductive traces on PDMS films, obtaining stretchable and compliant electrodes able to maintain their conductivity even under high mechanical strains [22]. With a size distribution range of 3–10 nm and a kinetic energy of about 0.5 eV atom, the clusters (made of several thousand atoms) have sufficient inertia to penetrate into the polymeric target, which is kept at room temperature (RT), and to form a nanocomposite layer, while avoiding charging and carbonization of the polymeric substrate [28]. The clusters implantation into the PDMS surface is evident by looking at the percolation curve monitored in situ during the deposition process (Figure 6.3a) and the transmission electron microscopy (TEM) images reported in Figure 6.3b. The performance of the conductive PDMS/

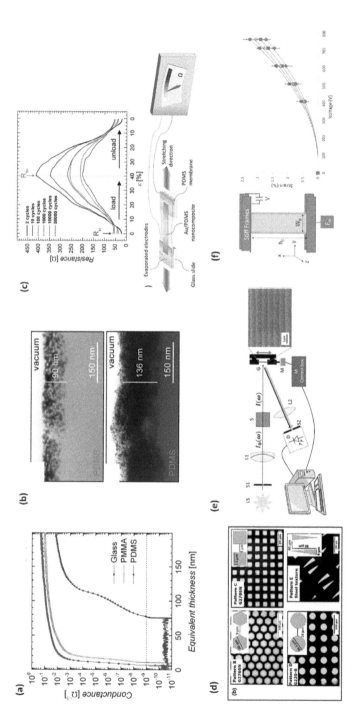

FIGURE 6.3 (a–c) Percolation curve, TEM pictures and electro-mechanical strain tests of the stretchable electrodes obtained by depositing Au clusters on the surface of PDMS by mean of SCBD (from Ref. [22]). (d) Micro-patterning of different geometries on PDMS using stencil masks (from Ref. [26]). (e) Scanning optical spectrometer based on an Ag-PDMS stretchable reflective grating (from Ref. [58]). (f) Dielectric elastomer actuators (DEA) obtained through the implantation of Ag clusters underneath PDMS surface (from Ref. [59]).

gold nanocomposites was tested against extensive uniaxial strain cycles (maximum tested strain of 40%), showing a remarkable behavior (Figure 6.3c). The nanocomposite resistance monotonically increased up to a value of ca. 400 Ω (R_{fin}), while it recovered its original value (R_{in}) during the unloading phase. The test was repeated 50,000 times and the resistance's variation trend remained basically unchanged, with only a slight increase in the value of R_{in} and a slight increase in R_{fin}. The samples were also tested up to their limit of mechanical failure (strain of 97%) and an electrical failure never occurred before the mechanical one. The observed long-term cyclic resiliency to the mechanical strain is due to the nanocomposite structure of the hybrid PDMS/Au layer, where the clusters are dispersed in the polymeric matrix forming a region whose mechanical properties are very close to that of the pristine PDMS. In fact, Atomic Force Microscopy (AFM) indentation technique was employed to study the mechanical properties of the PDMS/Au nanocomposite layers, and it was shown how they can be tuned accordingly to the Au volume concentration, up to values of Young's modulus 5 times higher than the pristine PDMS for an Au volume concentration of 30% [27].

SCBD showed to be not only an efficient method to produce stretchable and durable metallic circuits, but it was also demonstrated to allow the micro-patterning of different geometries by the use of stencil masks (Figure 6.3d), by exploiting the high collimation of the supersonic beam seeded with the metallic clusters. AFM technique was used to deeply investigate the morphology of the PDMS/Au patterns and the lateral resolution of the deposition technique, which was found to be of 0.5–1 μm [26]. The patterning of soft micro-electrodes on elastomeric materials was also obtained with the combination of SCBD and femtosecond laser ablation (FSLA) [56]. Initially, large areas of PDMS substrates have been additively provided by implanted cluster-assembled Au layers with a nominal thickness of 100 and 200 nm. Subsequently, the FSLA was employed as a subtractive technique for the selective ablation of the gold in a manner that only micrometric conductive path remained on the PDMS surface. This was possible due to the unique features that characterize the FSLA technique: high spatial precision, minimal collateral thermal damage, high material removal rate, selective ablation of thin layers and the capability of processing a large range of materials [57]. This approach allowed to obtain micro-electrodes with a width from 200 μm down to only 3 μm in every case their conductivity showed to vary in a linear way with their size.

SCBD technique represents a strategic tool of additive manufacturing for the implantation and the micro-patterning of conductive traces beneath the surface of elastomeric materials, leading to resilient and stretchable soft electrodes able to properly work under severe mechanical deformation. This approach has been exploited for the design and the development of different devices with interesting applications in optics, robotics and sensing [29,58,59]. A scanning optical spectrometer based on a PDMS stretchable reflective grating was produced by the implantation of Ag clusters (Figure 6.3e). The spectrometer was able to span the entire visible range of wavelengths without any moving mechanical component except the grating stretcher, and it maintained its optical performances for thousands of strain cycles [58]. The implantation of Ag nanoparticles in a PDMS substrate was also employed for the development of dielectric elastomer actuators (DEA) (Figure 6.3f), able to reach electrically induced strains up to 2.5% and with a reliable operability up to 10,000 cycles [59]. The custom-patterned implantation of Au clusters on both sides of a PDMS film generates an array of capacitive sensors able to independently detect the presence and the position of mechanically applied pressure [29]. Such a nano-composite touch pad was employed as a stretchable keyboard able to illuminate the lightning of a LED array according to which pad was touched.

6.4 METALLIZATION OF IONOGELS FOR ACTUATION, ENERGY HARVESTING AND STORAGE

Ionic liquids (ILs) are a class of electrolyte only composed of ions, with one of them at least of organic nature. Because of steric hindrance and a low lattice energy, they are in the liquid state at room temperature, and they possess unique features such as negligible vapor pressure, high ionic conductivity (10^{-4} to 10^{-2} S/cm), high viscosity and thermal, chemical and electrochemical high stability [60]. The incorporation of ILs into polymeric materials allows to obtain ionogels, a class of soft materials that combines the interesting physico-chemical properties of ILs and those of the polymer of choice. It is of particular interest to provide insulant polymeric materials with the IL's typical ionic conductivity. With the further integration of compliant and electronically conductive electrodes, hybrid nanocomposite materials can be developed where the application of a potential difference is associated with massive ionic accumulation and vice versa. Such soft and smart materials can find and even trigger novel applications in many fields such as robotics, soft actuation, sensors, energy harvesting, storage and management, environmental, and biomedical applications [61–63]. The deposition of supersonically accelerated Au clusters was exploited to form cluster-assembled conductive electrodes partially interpenetrating within the ionogels surface, leading to high compliancy and robust adhesion [33–48]. These hybrid Au/ionogel nanocomposite materials, integrating both ionic and electronic conductivity together with soft mechanical properties, allowed the development of several kinds of smart materials such as electro-mechanical soft actuators [64–67], piezoionic/piezoelectric systems [68] and supercapacitors [69,70].

A great deal of research has been focused on the development of electrically activated soft actuators [71–73]. In fact, soft actuation is a crucial aspect for the development of soft robotic systems and devices able to operate and adapt in unknown and/or extreme environments [74,75]. Electro-mechanical actuators are particularly interesting since electrical stimuli can be easily applied, controlled and interfaced with other devices [76], and properly designed metal/ionogel nanocomposites can be employed for this purpose. As can be seen in the schematics in Figure 6.4a, thin compliant electrodes need to be integrated on the opposite sides of the iono-gel film in a sandwich-like structure. Through the application of a potential difference among them, the ions migrate under the effect of the imposed electric field and they accumulate near the electrode of opposite charge. If the (solvated) positive and negative ions are characterized by a different size, a heterogeneous redistribution of the IL occurs inside the ionogel, causing an asymmetric swelling/shrinking that leads to an overall bending of the soft actuator. Such electro-mechanical actuation is reversible, bidirectional and easy to control by mean of electric signals. In this

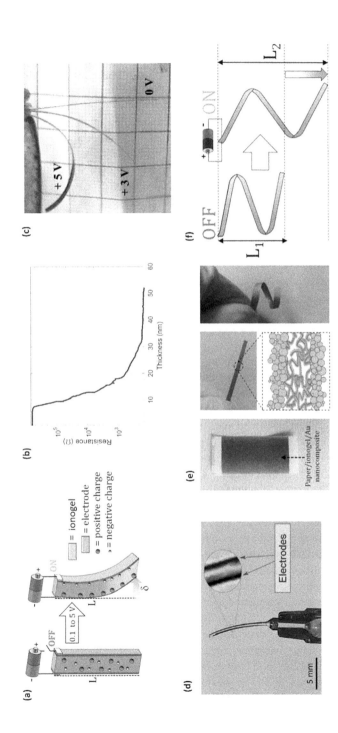

FIGURE 6.4 (a) Operational principle of ionogel/metal electro-mechanical soft actuators (from Ref. [33]). (b and c) Percolation curve and electro-mechanical actuation tests of the poly(acrylic acid-co-acrylonitrile)/(EMIM)BF$_4$/Au soft actuators (from Ref. [64]). (d) Cylindrical soft ionogel/Au actuators (from Ref. [67]). (e) Paper strips impregnated with electroactive ionogel and provided with cluster-assembled Au electrodes (from Ref. [65]). (f) Schematics of the spring-like electro-mechanical motion of the paper/ionogel/Au soft actuators (from Ref. [65]).

context, the production of well-adherent, highly conductive and thin electrodes able to undergo stable cyclic deformations along with the electroactive ionogels is maybe the most important and a critical task. The use of SCBD allowed to provide compliant and resilient cluster-assembled electrodes to different kinds of ionogels to obtain electro-mechanical soft actuators. An ionogel thin film based on a poly(acrylic acid-co-acrylonitrile) polymeric matrix and the IL 1-ethyl-3-methylimidazolium tetrafluoroborate (EMIM-BF$_4$) was obtained with a photo-polymerization reaction. The organic salt tetraethylammonium fluoride (TEAF) and halloysite nanoclays (HNCs) were also incorporated to improve the asymmetry of the ionic migration inside the gel. 100-nm-thick compliant gold electrodes were provided by means of SCBD and the corresponding percolation curve is reported in Figure 6.4b. The obtained soft actuators showed a high electro-sensitivity down to an applied potential of 0.1 V, high performance and reversible actuation in a low-voltage range between 2 and 5 V, as well as good durability in their frequency response (up to 76,000 cycles at 2 V and 1 Hz) (Figure 6.4c) [64]. A similar approach was combined with replica molding in order to manufacture cylindrical three-dimensional micro-actuators with high aspect ratios (Figure 6.4d) [67].

Plain paper was also employed as a porous backbone by soaking in the pre-polymer solution and subsequent photo-polymerization of the ionogel imbibed in the paper's cellulose fibers. This strategy allowed folding to achieve a spring-like morphology (Figure 6.4e and f) able to provide longitudinal electro-mechanical actuation with good stability up to 5,000 cycles [65]. Not less importantly, 74% (in weight) of the overall smart composite material was represented by natural-derived and biodegradable paper. Another step toward the development of eco-friendly actuators was the use of SCBD to metallize a different kind of ionogel, obtained from solvent casting by blending together the natural-derived bioplastic poly(3-hydroxybutyrrate) (PHB) and the IL 1-Butyl-3-methylimidazoliumbis-(trifluoromethylsulfonyl)imide (BMIM-TFSI) [66]. The so obtained ionogel was based on hydrophobic building blocks and it resulted to be completely resistant to water and unaffected by the surrounding humidity, making him an ideal candidate to work in unknown or evolving environments. The electro-mechanical actuation performances were comparable with those of traditional non-biodegradable actuators, but with the great advantage to be unaffected by the surrounding humidity level and to be derived from a biopolymer.

Metal/ionogels nanocomposite films, thanks to their hybrid electronic-ionic conduction, can act as electro-mechanical actuators, but they can also be employed to convert mechanical forces into an electric field. In fact, the so-called piezoionic ionogels possess the capability to generate an output voltage induced by the separation of ions with different mobility, stimulated by a differential pressure applied to the material [77–79]. They are usually responsive to stimuli at low frequencies (<1 Hz) providing output voltages of ca. 1–10 mV [80,81]. Thanks to this, piezoionic ionogels can be used to carry out energy harvesting and sensing functionalities [82–87]. Efficient energy conversion, flexibility and stretchability must be shared by both the ionogel and the coupled electrodes. SCBD has been employed to produce Au compliant electrodes on the opposite side of a piezoionic ionogel, composed of a polymeric matrix of   poly(2-(hydroxymethyl)methacrylate-co-acrylonitrile)   and   polyvinylpyrrolidone (PVP) and the IL 1-(2-hydroxyethyl)-3-methylimidazolium tetrafluoroborate

(HOEMIM-BF$_4$). Piezoelectric nanoparticles of barium titanate (BaTiO$_3$) were also embedded into the ionogel to obtain a nanocomposite material combining the low frequency sensitivity, typical of piezoionics, with the anisotropic output voltage typical of piezoelectrics (Figure 6.5a). The deposition of compliant cluster-assembled Au electrodes by means of SCBD resulted in an enhancement of the charge accumulation efficiency by a factor of ten, compared to the same tests carried out with bulk metal electrodes physically put in contact with the ionogel films [68]. This is due to the tight and nanostructured interface formed by the partial implantation of the Au clusters into the ionogels surface.

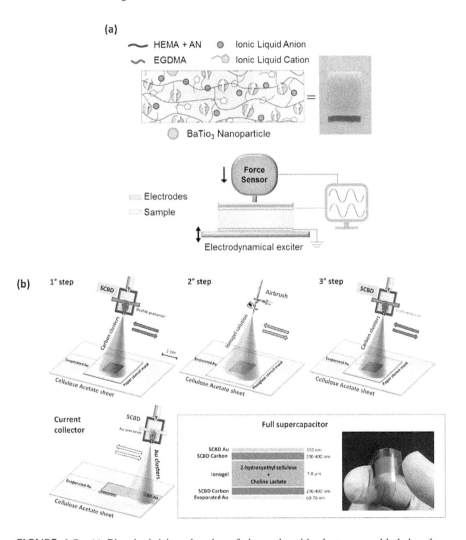

FIGURE 6.5 (a) Piezoionic/piezoelectric soft ionogels with cluster-assembled Au electrodes (from Ref. [68]). (b) Layer-by-layer all-printed fabrication process of biodegradable micro-supercapacitors, obtained by the coupling of aqueous spray casting and supersonic cluster beam deposition (from Ref. [70]).

Besides the coupling between electrical and mechanical stimuli, the accumulation of charged ions at the ionogel-electrode interface has also been exploited to obtain electrolytic double-layer capacitors (EDLCs), a particular kind of the so-called supercapacitors (SCs) [88–90]. In fact, by the application of a potential difference between two electrodes on the opposite side of an ionogel, it is possible to reversibly store electrical energy in the formation of an electrolytic double layer at the ionogel-electrode interfaces. Compared to other energy storage devices like batteries, SCs are characterized by higher power and cyclic stability (up to 10^6 cycles). Unlike batteries, which can be charged only at fixed potential differences, supercapacitors can be charged even by low-voltage devices that harvest green energy from the surrounding environment, such as biofuel cells, photovoltaics, piezoelectric or triboelectric materials [91–93]. By taking into account their low structural complexity and the possibility to employ natural-derived and biodegradable materials as polymeric backbones, SCs are ideal candidates for energy storage in wearable devices and miniature environmental sensors [94,95]. In this kind of devices, the nature and the structure of the ionogel-electrode interface are of crucial importance. In fact, the amount of stored energy is directly proportional to the number of ions that can be accumulated, which in turn depends on the surface area of the electrodes. Highly porous and conductive compliant electrodes are then of primary importance for the development of high power and high energy soft supercapacitors, and the technique of supersonic cluster beam deposition has been successfully employed for this purpose. Eco-friendly, flexible and biodegradable micro-supercapacitors were obtained by combining natural polymers, such as cellulose and bioplastics, with green ILs and compliant gold and carbon electrodes by means of SCBD [69,70]. The electrodes cluster-assembled structure, as well as their partial implantation underneath the ionogels surface, formed a highly porous high-surface area interface that helped the formation of an electrolytic double layer where a high number of charged ions could be accumulated (maximum double-layer capacitance of 1 mF/cm^2). This approach allowed to obtain ultrathin (less than 10 μm) biodegradable supercapacitors able to store energy amounts up to 12 Wh/kg and with maximum delivery powers of 120 kW/kg. The coupling of SCBD with a spray casting technique resulted in an all-printed fabrication process suitable for scalability, automation and micro-patterning (Figure 6.5b).

6.5 CONDUCTIVE TRACES ON 3D-PRINTED OBJECTS: 3D-PRINTED ELECTRONICS

AM refers to any technology relying on the fabrication of an object by joining materials, usually layer upon layer, directly from its three-dimensional virtual model [15]. These approaches enable the fabrication of 3D freeform components with complex geometries from a variety of materials (e.g. metals, ceramics, polymers, nanocomposites) [15–17] and, among them, 3D printing (3DP) is currently the most widespread and extensively used for rapid prototyping, i.e., for quick production of scale models of a part or finished product [18]. Recent advances in 3DP technologies in terms of hardware and software development suggest that they will be increasingly used in industry for small-scale production and they are expected to reach a sufficient maturity for end products manufacturing and mass production by 2050 [96].

Recently, great effort has been focused in the development and fabrication of smart objects and functional devices with integrated sensing and actuation capabilities able to interact with the surrounding environment, leading to a push of AM technologies toward multi-material printing and printed electronics [97,98]. 3D printed electronics relies on the concept of integrating conductive structures, wires, passive electric and active electronic components into 3D printed monolithic platforms, in order to confer the desired functional behavior to the object in a single (or few steps) fabrication process. The 3DP of smart electronic objects is predominantly based on the production of conductive parts employing nanocomposite functional inks, usually constituted by a liquid mixture containing high loadings of metal nanoparticles (usually silver or copper) dispersed in a suitable solvent. Such inks are usually directly "written" on 3DP objects reproducing the 2D design and pattern of specific circuitries in an automated fashion. For example, an effective route to process conductive inks for 3D printed electronics is the AJP technique [99–101]. Nevertheless, some issues remained problematic for this kind of approach: the relatively low spatial resolution (minimum feature size achievable is around few micrometers), the poor control over the electrical resistance of the printed structures, the use of post-process thermal treatments that can limit the range of substrate materials, the high particles loading of the ink (>70% of the total weight), the adhesion between the conductive layer and the base polymer.

SCBD can represent an alternative technique for the printing of electronics onto 3DP objects of various nature. In fact, it is the technique compatible with planar fabrication technologies, and it enables the patterning of highly resolved features using stencil masks, with lateral resolution down to one micron and no shadow effects due to the low beam divergence. Due to their low kinetic energies, the metallic clusters can penetrate polymeric substrates down to a characteristic depth and to progressively build up the surface of the substrate to form a well-adhering conductive nanocomposite film, with electrical resistance depending only on the number of nanoparticles deposited. Thanks to these features, SCBD can be considered an additive manufacturing technique, since it enables the printing of conductive structures onto polymers by additively generating cluster-assembled layers, joining nanoparticles together in a "pulse-by-pulse" fashion and preserving their structure and functionality upon deposition [29]. The combination of SCBD with the AM technique of FFF [102] constituted an integrated additive manufacturing approach consisting in a multi-step procedure, alternating the two techniques to generate a 3D plastic object with integrated polymer/metal nanocomposite structures, such as 3D conductive lines, resistors with tailored electrical conductivity, 3D bridge vias and sockets for standard electronic components fitting. It allowed the design and the fabrication of a microfluidic electrochemical biosensor [30] as well as the prototyping of an all-printed monolithic dark-activated light sensor [31].

Electrochemical microfluidic biosensors have emerged as powerful tools for high sensitivity chemical and biological detection and analysis of a wide range of compounds, from synthetic pesticides and hazardous chemicals to metabolites and cell-secreted biomolecules [103]. They are usually based on an electrochemical detection unit incorporated in a microfluidic platform. The synergy between control over fluids in miniaturized polymeric environments and electrochemical detection at metallic thin films electrodes enables the coupling of enhanced accuracy and selectivity with disposability, reduced cost, portability, real-time and high-throughput detection capability [104]. Because of these advantages, such devices retain huge potential in many fields such as health care [105], water and food quality assurances [106], agriculture [107] and environmental monitoring [108]. The technological challenge is constituted by the development of suitable fabrication approaches for the integration of electrochemically stable metal electrode into polymer-based microfluidic platforms. To address this challenge, SCBD technique has been employed together with FFF technique and selective electroplating for the additive manufacturing of a microfluidic electrochemical sensor provided with integrated functional electrodes, operating as a classical three-electrode electrochemical cell [30]. The design of the device is shown in Figure 6.6a.

A three-electrode electrochemical cell was meant to be placed into the microfluidic channel of the platform made of acrylonitrile-butadiene-styrene (ABS), 3D printed with FFF technique and provided with a millimeter-size fluidic channel. SCBD technique was employed to print three gold electrodes inside the channel, together with three conductive lines leading to external interconnection pads. The desired geometries could be obtained using custom-shaped stencil masks. Two of the Au electrodes were employed as working and counter electrodes, while the third was selectively electro-plated with silver/silver chloride to work as a semi-reference electrode. An ABS top layer was then overprinted to seal the internal channel, leaving two through-holes (4 mm diameter) to serve as fluidic ports, fitting commercial tubing connectors to connect the device with an external syringe pump. The operability of electrochemical sensor was assessed through standard electrochemical testing in both stagnant and laminar flow regimes, carried out on aqueous solution containing the redox couple potassium ferricyanide/ferrocyanide in potassium nitrate as a probe (Figure 6.6b). The electrochemical response of the device assessed in dynamic conditions suggests how the combination of FFF, SCBD and selective electroplating can be used to manufacture continuous flow microfluidics-based functional systems for electrochemical detection on-chip and/or on non-conventional substrates.

The merging of FFF and SCBD can also offer new opportunities in printing 3D electronic circuits with simplified topology, enabling inter- and intra-layer connections between electrical components with conducting vias and 3D bridging. Moreover, the presence of highly conductive socket slots avoids the use of soldering or conductive pastes to implement bulky electrical and electronic elements. This approach allowed to prototype and to validate a dark-activated light sensor [31]. The device relied on the use of a photo-resistor with a characteristic resistance (R_f) that increases when the ambient light intensity drops below a certain threshold. The corresponding drop of the voltage imposed across the sensing element (V_{sens}) determines the switching-on of one or two light-emitting diodes (LEDs) whether V_{sens} falls below one or two fixed reference voltage signals $(V_{ref1}$ and $V_{ref2})$, respectively. A 4.5 V power supply, two voltage dividers, two operational amplifiers and two LEDs were integrated in the device to allow its proper operation. The circuit layout was translated into the printed smart object as schematized in Figure 6.6c and d. The architecture consists of three distinct levels, interconnected through SCBD printed oblique conductive paths: (1) a battery box for power supply, able to host commercial batteries; (2) a circuit board provided

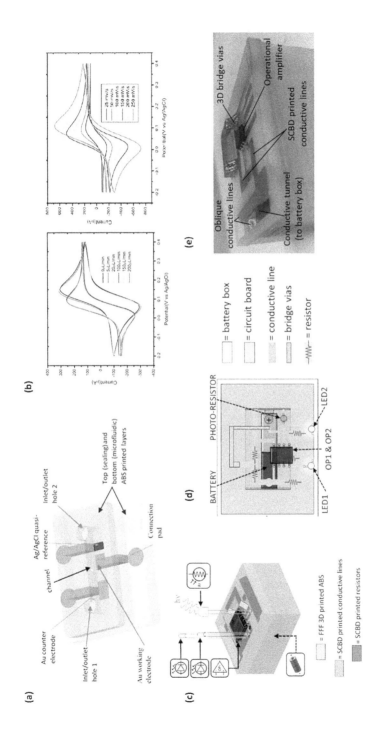

FIGURE 6.6 (a) Design of the microfluidic biosensor obtained through the combination of FFF and SCBD techniques (from Ref. [30]). (b) Biosensors cyclic voltammetry carried out by varying the flow regime and the applied potential scan rate (from Ref. [30]). (c–e) Design and photograph of a monolithic dark-activated light sensor obtained by the merging FFF and SCBD technique (adapted from Refs. [29,31]).

with SCBD conductive lines and resistors, sockets and active electronic components; and (3) a 3D bridge via for out-of-plane connections of the circuit board elements. SCBD was sequentially combined with FFF to print functional structures onto ABS. Resistors with electrical resistance ranging from around 10 GΩ to 1 Ω were deposited with SCBD and their resistance was tailored accordingly to the number of nanoparticles deposited. A photograph of the complete device is shown in Figure 6.6e. Its functionality was assessed by exposing the system to three different illumination conditions (dark, ambient light and a 150 W light bulb) and by monitoring in real time the corresponding voltage response. With these results, the integration of FFF and SCBD into a single manufacturing process showed to be an effective solution for the additive fabrication of monolithic devices with embedded functional structures.

6.6 CONCLUSION

We have discussed some examples of the application of SCBD for the printing of electrically conductive patterns on systems and devices based on soft polymers and 3D printed substrates. This fabrication method preserves the structural and functional properties after metallization due to the cluster-assembled nature of the nanostructured electrodes. The presented approach enables the high-throughput fabrication of a novel class of hybrid systems with complex geometries and properties. We foresee that the use of SCBD constitutes a promising technology for the development and fabrication of resilient polymer-based devices in view of the realization of bio-inspired soft robots and robotic architectures, smart sensors, and resilient architectures deployable in different environments.

REFERENCES

1. Rich SI, Wood RJ, Majidi C. Untethered soft robotics. *Nat Electron.* 2018;1(2):102–12.
2. Valentine AD, Busbee TA, Boley JW, Raney JR, Chortos A, Kotikian A, et al. Hybrid 3D printing of soft electronics. *Adv Mater.* 2017;29(40):1–8.
3. Wang H, Totaro M, Beccai L. Toward perceptive soft robots: progress and challenges. *Adv Sci.* 2018;5(9):1800541.
4. Koshima H. Mechanically responsive materials for soft robotics. Mechanically Responsive Materials for Soft Robotics. Wiley-VCH Verlag, New York, 2020.
5. Lu N, Kim DH. Flexible and stretchable electronics paving the way for soft robotics. *Soft Robot.* 2014;1(1):53–62.
6. Sachyani Keneth E, Kamyshny A, Totaro M, Beccai L, Magdassi S. 3D printing materials for soft robotics. *Adv Mater.* 2021;33(19):2003387.
7. Majidi C. Soft-matter engineering for soft robotics. *Adv Mater Technol.* 2019;4(2):1800477.
8. Gao W, Emaminejad S, Nyein HYY, Challa S, Chen K, Peck A, et al. Fully integrated wearable sensor arrays for multiplexed in situ perspiration analysis. *Nature.* 2016;529(7587):509–14.
9. Runciman M, Darzi A, Mylonas GP. Soft Robotics in minimally invasive surgery. *Soft Robot.* 2019;6(4):423–43.
10. Xiang L, Xia F, Zhang H, Liu Y, Liu F, Lia ng X, et al. Wafer-scale high-yield manufacturing of degradable electronics for environmental monitoring. *Adv Funct Mater.* 2019;29(50):1–9.

11. Chowdhary G, Gazzola M, Krishnan G, Soman C, Lovell S. Soft robotics as an enabling technology for agroforestry practice and research. *Sustain.* 2019;11(23):6751.
12. Li R, Wang L, Kong D, Yin L. Recent progress on biodegradable materials and transient electronics. *Bioact Mater.* 2018;3(3):322–33.
13. Rossiter J, Winfield J, Ieropoulos I. Here today, gone tomorrow: biodegradable soft robots. *Electroact Polym Actuators Devices.* 2016. 2016;9798:97981S.
14. Akogwu O, Kwabi D, Midturi S, Eleruja M, Babatope B, Soboyejo WO. Large strain deformation and cracking of nano-scale gold films on PDMS substrate. *Mater Sci Eng B Solid-State Mater Adv Technol.* 2010 Jun;170(1–3):32–40.
15. Tofail SAM, Koumoulos EP, Bandyopadhyay A, Bose S, O'Donoghue L, Charitidis C. Additive manufacturing: scientific and technological challenges, market uptake and opportunities. *Mater Today.* 2018;21(1):22–37.
16. Horn TJ, Harrysson OLA. Overview of current additive manufacturing technologies and selected applications. *Sci Prog.* 2012;95(3):255–82.
17. Singh S, Ramakrishna S, Singh R. Material issues in additive manufacturing: a review. *J Manuf Process.* 2017;25:185–200.
18. Karania R, Kazmer D. Low volume plastics manufacturing strategies. *J Mech Des.* 2007;129(12):1225.
19. Piseri P, Tafreshi HV, Milani P. Manipulation of nanoparticles in supersonic beams for the production of nanostructured materials. *Curr Opin Solid State Mater Sci.* 2004;8(3–4):195–202.
20. Wegner K, Piseri P, Tafreshi HV, Milani P. Cluster beam deposition: a tool for nanoscale science and technology. *J Phys D Appl Phys.* 2006;39(22):R439.
21. Barborini E, Vinati S, Leccardi M, Repetto P, Bertolini G, Rorato O, et al. Batch fabrication of metal oxide sensors on micro-hotplates. *J Micromech Microeng.* 2008;18(5):055015.
22. Corbelli G, Ghisleri C, Marelli M, Milani P, Ravagnan L. Highly deformable nanostructured elastomeric electrodes with improving conductivity upon cyclical stretching. *Adv Mater.* 2011;23(39):4504–8.
23. Barborini E, Piseri P, Podesta' A, Milani P. Cluster beam microfabrication of patterns of three-dimensional nanostructured objects. *Appl Phys Lett.* 2000;77(7):1059–61.
24. Caruso F, Bellacicca A, Milani P. High-throughput shadow mask printing of passive electrical components on paper by supersonic cluster beam deposition. *Appl Phys Lett.* 2016;108(16):163501.
25. Cavaliere E, De Cesari S, Landini G, Riccobono E, Pallecchi L, Rossolini GM, et al. Highly bactericidal Ag nanoparticle films obtained by cluster beam deposition. *Nanomed Nanotechnol Biol Med.* 2015;11(6):1417–23.
26. Ghisleri C, Borghi F, Ravagnan L, Podestà A, Melis C, Colombo L, et al. Patterning of gold-polydimethylsiloxane (Au-PDMS) nanocomposites by supersonic cluster beam implantation. *J Phys D Appl Phys.* 2014;47(1):015301.
27. Borghi F, Melis C, Ghisleri C, Podestà A, Ravagnan L, Colombo L, et al. Stretchable nanocomposite electrodes with tunable mechanical properties by supersonic cluster beam implantation in elastomers. *Appl Phys Lett.* 2015;106(12):121902.
28. Ravagnan L, Divitini G, Rebasti S, Marelli M, Piseri P, Milani P. Poly(methyl methacrylate)-palladium clusters nanocomposite formation by supersonic cluster beam deposition: a method for microstructured metallization of polymer surfaces. *J Phys D Appl Phys.* 2009;42(8):082002.
29. Santaniello T, Milani P. Additive nano-manufacturing of 3D printed electronics using supersonic cluster beam deposition. *Front Nanosci.* 2020;15:313–33.
30. Gebreyes WA, Migliorini L, Pezzotta F, Shacham-Diamand Y, Santaniello T, Milani P. An integrated fluidic electrochemical sensor manufactured using fused filament fabrication and supersonic cluster beam deposition. *Sens Actuators A Phys.* 2020;301:111706.

31. Bellacicca A, Santaniello T, Milani P. Embedding electronics in 3D printed structures by combining fused filament fabrication and supersonic cluster beam deposition. *Addit Manuf.* 2018;24:60–6.

32. Wilkinson NJ, Smith MAA, Kay RW, Harris RA. A review of aerosol jet printing-a non-traditional hybrid process for micro-manufacturing. *Int J Adv Manuf Technol.* 2019;105(11):4599–619.

33. Santaniello T, Migliorini L, Yan Y, Lenardi C, Milani P. Supersonic cluster beam fabrication of metal – ionogel nanocomposites for soft robotics. *J Nanoparticle Res.* 2018;20(9):1–19.

34. Xie LH, Ling QD, Hou XY, Huang W. An effective Friedel-Crafts postfunctionalization of poly(N-vinylcarbazole) to tune carrier transportation of supramolecular organic semiconductors based on π-stacked polymers for nonvolatile flash memory cell. *J Am Chem Soc.* 2008;130(7):2120–1.

35. Zhang YZ, Wang Y, Cheng T, Lai WY, Pang H, Huang W. Flexible supercapacitors based on paper substrates: a new paradigm for low-cost energy storage. *Chem Soc Rev.* 2015;44(15):5181–99.

36. An Z, Zheng C, Tao Y, Chen R, Shi H, Chen T, et al. Stabilizing triplet excited states for ultralong organic phosphorescence. *Nat Mater.* 2015;14(7):685–90.

37. Jiang J, Zhu J, Ai W, Wang X, Wang Y, Zou C, et al. Encapsulation of sulfur with thin-layered nickel-based hydroxides for long-cyclic lithium-sulfur cells. *Nat Commun.* 2015;6(1):1–9.

38. Angmo D, Andersen TR, Bentzen JJ, Helgesen M, Søndergaard RR, Jørgensen M, et al. Roll-to-roll printed silver nanowire semitransparent electrodes for fully ambient solution-processed tandem polymer solar cells. *Adv Funct Mater.* 2015;25(28):4539–47.

39. Zhong Z, Lee H, Kang D, Kwon S, Choi YM, Kim I, et al. Continuous patterning of copper nanowire-based transparent conducting electrodes for use in flexible electronic applications. *ACS Nano.* 2016;10(8):7847–54.

40. Du J, Pei S, Ma L, Cheng HM. 25th anniversary article: carbon nanotube- and graphene-based transparent conductive films for optoelectronic devices. *Adv Mater.* 2014;26(13):1958–91.

41. Ellmer K. Past achievements and future challenges in the development of optically transparent electrodes. *Nat Photonics.* 2012;6(12):809–17.

42. Hecht DS, Hu L, Irvin G. Emerging transparent electrodes based on thin films of carbon nanotubes, graphene, and metallic nanostructures. *Adv Mater.* 2011;23(13):1482–513.

43. Cheng T, Zhang Y, Lai WY, Huang W. Stretchable thin-film electrodes for flexible electronics with high deformability and stretchability. *Adv Mater.* 2015;27(22):3349–76.

44. Amjadi M, Pichitpajongkit A, Lee S, Ryu S, Park I. Highly stretchable and sensitive strain sensor based on silver nanowire-elastomer nanocomposite. *ACS Nano.* 2014;8(15):5154–63.

45. Roh E, Hwang BU, Kim D, Kim BY, Lee NE. Stretchable, transparent, ultrasensitive, and patchable strain sensor for human-machine interfaces comprising a nanohybrid of carbon nanotubes and conductive elastomers. *ACS Nano.* 2015;9(6):6252–61.

46. Yao S, Zhu Y. Wearable multifunctional sensors using printed stretchable conductors made of silver nanowires. *Nanoscale.* 2014;6(4):2345–52.

47. Kang D, Pikhitsa P V., Choi Y W, Lee C, Shin SS, Piao L, et al. Ultrasensitive mechanical crack-based sensor inspired by the spider sensory system. *Nature.* 2014;516(7530):222–6.

48. Zheng M, Li W, Xu M, Xu N, Chen P, Han M, et al. Strain sensors based on chromium nanoparticle arrays. *Nanoscale.* 2014;6(8):3930–3.

49. Migliorini L. *Development of Functional Nanocomposite Materials Towards Biodegradable Soft Robotics and Flexible Electronics.* Ph Dissertation, University of Milan, Dept of Chemistry, 2020. https://air.unimi.it/retrieve/dfa8b9a0-93eb-748b-e053-3a05fe0a3a96/phd_unimi_R11729.pdf

50. McCoul D, Hu W, Gao M, Mehta V, Pei Q. Recent advances in stretchable and transparent electronic materials. *Adv Electron Mater.* 2016;2(5):1500407.
51. Liu Y, Pharr M, Salvatore GA. Lab-on-skin: a review of flexible and stretchable electronics for wearable health monitoring. *ACS Nano.* 2017;11(10):9614–35.
52. Whitesides GM. The origins and the future of microfluidics. *Nature.* 2006;442(7101):368–73.
53. Graz IM, Cotton DPJ, Lacour SP. Extended cyclic uniaxial loading of stretchable gold thin-films on elastomeric substrates. *Appl Phys Lett.* 2009;94(7):071902.
54. Rosset S, Niklaus M, Dubois P, Shea HR. Metal ion implantation for the fabrication of stretchable electrodes on elastomers. *Adv Funct Mater.* 2009;19(3):470–8.
55. Maggioni G, Vomiero A, Carturan S, Scian C, Mattei G, Bazzan M, et al. Structure and optical properties of Au-polyimide nanocomposite films prepared by ion implantation. *Appl Phys Lett.* 2004;85(23):5712–4.
56. Dotan T, Berg Y, Migliorini L, Villa SM, Santaniello T, Milani P, et al. Soft and flexible gold microelectrodes by supersonic cluster beam deposition and femtosecond laser processing. *Microelectron Eng.* 2021;237:111478.
57. Chen JK, Beraun JE. Modelling of ultrashort laser ablation of gold films in vacuum. *J Opt A Pure Appl Opt.* 2003;5(3):168–73.
58. Ghisleri C, Potenza MAC, Ravagnan L, Bellacicca A, Milani P. A simple scanning spectrometer based on a stretchable elastomeric reflective grating. *Appl Phys Lett.* 2014;104(6):061910.
59. Taccola S, Bellacicca A, Milani P, Beccai L, Greco F. Low-voltage dielectric elastomer actuators with stretchable electrodes fabricated by supersonic cluster beam implantation. *J Appl Phys.* 2018;124(6):064901.
60. Zhang S, Sun N, He X, Lu X, Zhang X. Physical properties of ionic liquids: database and evaluation. *J Phys Chem Ref Data.* 2006;35(4):1475–517.
61. Correia DM, Fernandes LC, Martins PM, García-Astrain C, Costa CM, Reguera J, et al. Ionic liquid-polymer composites: a new platform for multifunctional applications. *Adv Funct Mater.* 2020;30(24):1909736.
62. Le Bideau J, Viau L, Vioux A. Ionogels, ionic liquid based hybrid materials. *Chem Soc Rev.* 2011;40(2):907–25.
63. Andrzejewska E, Marcinkowska A, Zgrzeba A. Ionogels – materials containing immobilized ionic liquids. *Polimery/Polymers.* 2017;62(5):344–52.
64. Yan Y, Santaniello T, Bettini LG, Minnai C, Bellacicca A, Porotti R, et al. Electroactive ionic soft actuators with monolithically integrated gold nanocomposite electrodes. *Adv Mater.* 2017;29(23):1606109.
65. Santaniello T, Migliorini L, Borghi F, Yan Y, Rondinini S, Lenardi C, et al. Spring-like electroactive actuators based on paper/ionogel/metal nanocomposites. *Smart Mater Struct.* 2018;27(6):065004.
66. Migliorini L, Santaniello T, Rondinini S, Saettone P, Comes Franchini M, Lenardi C, et al. Bioplastic electromechanical actuators based on biodegradable poly(3-hydroxybutyrate) and cluster-assembled gold electrodes. *Sens Actuators B Chem.* 2019;286:230–6.
67. Milana E, Santaniello T, Azzini P, Migliorini L, Milani P. Fabrication of high-aspect-ratio cylindrical micro-structures based on electroactive ionogel/gold nanocomposite. *Appl Nano.* 2020;1(1):59–69.
68. Villa SM, Mazzola VM, Santaniello T, Locatelli E, Maturi M, Migliorini L, et al. Soft piezoionic/piezoelectric nanocomposites based on ionogel/batio 3 nanoparticles for low frequency and directional discriminative pressure sensing. *ACS Macro Lett.* 2019;8(4):414–20.
69. Migliorini L, Santaniello T, Borghi F, Saettone P, Franchini MC, Generali G, et al. Eco-friendly supercapacitors based on biodegradable poly(3-hydroxy-butyrate) and ionic liquids. *Nanomaterials.* 2020;10(10):2062.

70. Migliorini L, Piazzoni C, Põhako-Esko K, DI Girolamo M, Vitaloni A, Borghi F, et al. All-printed green micro-supercapacitors based on a natural-derived ionic liquid for flexible transient electronics. *Adv Funct Mater.* 2021;31:2102180.
71. Rinne P, Põldsalu I, Ratas HK, Kruusamäe K, Johanson U, Tamm T, et al. Fabrication of carbon-based ionic electromechanically active soft actuators. *J Vis Exp.* 2020;158:e61216.
72. Mirfakhrai T, Madden JDW, Baughman RH. Polymer artificial muscles. *Mater Today* 2007;10(4):30–8.
73. Bar-Cohen Y. Electroactive polymers as artificial muscles: a review. *J Spacecr Rockets.* 2002;39(6):822–7.
74. Rus D, Tolley MT. Design, fabrication and control of soft robots. *Nature.* 2015;521(7553):467–75.
75. Laschi C, Cianchetti M. Soft robotics: new perspectives for robot bodyware and control. *Front Bioeng Biotechnol.* 2014;2:3.
76. Martins P, Correia DM, Correia V, Lanceros-Mendez S. Polymer-based actuators: back to the future. *Phys Chem Phys.* 2020;22(27):15163–82.
77. Zhao J, Han S, Yang Y, Fu R, Ming Y, Lu C, et al. Passive and space-discriminative ionic sensors based on durable nanocomposite electrodes toward sign language recognition. *ACS Nano.* 2017;11(9):8590–9.
78. Triandafilidi V, Hatzikiriakos SG, Rottler J. Molecular simulations of the piezoionic effect. *Soft Matter.* 2018;14(30):6222–9.
79. Sarwar MS, Dobashi Y, Glitz E, Farajollahi M, Mirabbasi S, Naficy S, et al. Transparent and conformal "piezoionic" touch sensor. In: *Electroactive Polymer Actuators and Devices (EAPAD) 2015.* 2015. p. 943026.
80. Dobashi Y, Allegretto G, Sarwar MS, Cretu E, Madden JDW. Mechanoionic transduction of solid polymer electrolytes and potential applications. *MRS Adv.* 2016;1(1):63–8.
81. Liu Y, Hu Y, Zhao J, Wu G, Tao X, Chen W. Self-powered piezoionic strain sensor toward the monitoring of human activities. *Small.* 2016;12(36):5074–80.
82. Futaba DN, Hayamizu Y, Hata K, Yamamoto Y, Yamada T, Izadi-Najafabadi A, et al. A stretchable carbon nanotube strain sensor for human-motion detection. *Nat Nanotechnol.* 2011;6(5):296–301.
83. Cornogolub A, Cottinet PJ, Petit L. Hybrid energy harvesting systems, using piezoelectric elements and dielectric polymers. *Smart Mater Struct.* 2016;25(9):095048.
84. Prateek, Thakur VK, Gupta RK. Recent progress on ferroelectric polymer-based nanocomposites for high energy density capacitors: synthesis, dielectric properties, and future aspects. *Chem Rev.* 2016;116(7):4260–317.
85. Chen X, Parida K, Wang J, Xiong J, Lin MF, Shao J, et al. A stretchable and transparent nanocomposite nanogenerator for self-powered physiological monitoring. *ACS Appl Mater Interfaces.* 2017;9(48):42200–9.
86. Liu X, Wu D, Wang H, Wang Q. Self-recovering tough gel electrolyte with adjustable supercapacitor performance. *Adv Mater.* 2014;26(25):4370–5.
87. Zhang S, Wang F, Peng H, Yan J, Pan G. Flexible highly sensitive pressure sensor based on ionic liquid gel film. *ACS Omega.* 2018;3(3):3014–21.
88. Vangari M, Pryor T, Jiang L. Supercapacitors: review of materials and fabrication methods. *J Energy Eng.* 2013;139(2):72–9.
89. González A, Goikolea E, Barrena JA, Mysyk R. Review on supercapacitors: technologies and materials. *Renew Sustain Energy Rev.* 2016;58:1189–206.
90. Zhang L, Hu X, Wang Z, Sun F, Dorrell DG. A review of supercapacitor modeling, estimation, and applications: a control/management perspective. *Renew Sustain Energy Rev.* 2018;81:1868–78.

91. Liu R, Takakuwa M, Li A, Inoue D, Hashizume D, Yu K, et al. An efficient ultra-flexible photo-charging system integrating organic photovoltaics and supercapacitors. *Adv Energy Mater.* 2020;10(20):1–8.

92. Dong K, Wang YC, Deng J, Dai Y, Zhang SL, Zou H, et al. A highly stretchable and washable all-yarn-based self-charging knitting power textile composed of fiber triboelectric nanogenerators and supercapacitors. *ACS Nano.* 2017;11(9):9490–9.

93. Shen F, Pankratov D, Pankratova G, Toscano MD, Zhang J, Ulstrup J, et al. Supercapacitor/biofuel cell hybrid device employing biomolecules for energy conversion and charge storage. *Bioelectrochemistry.* 2019;128:94–9.

94. Huang S, Zhu X, Sarkar S, Zhao Y. Challenges and opportunities for supercapacitors. *APL Mater.* 2019;7(10):100901.

95. Dubal DP, Chodankar NR, Kim DH, Gomez-Romero P. Towards flexible solid-state supercapacitors for smart and wearable electronics. *Chem Soc Rev.* 2018;47(6):2065–129.

96. Cotteleer MJ. 3D opportunity: additive manufacturing paths to performance, innovation, and growth. *SIMT Addit Manuf Symp.* 2014;14:5–19.

97. MacDonald E, Salas R, Espalin D, Perez M, Aguilera E, Muse D, et al. 3D printing for the rapid prototyping of structural electronics. *IEEE Access.* 2014;2:234–42.

98. Espalin D, Muse DW, MacDonald E, Wicker RB. 3D Printing multifunctionality: Structures with electronics. *Int J Adv Manuf Technol.* 2014;72(5–8):963–78.

99. Gupta AA, Bolduc A, Cloutier SG, Izquierdo R. Aerosol Jet Printing for printed electronics rapid prototyping. In: *Proceedings – IEEE International Symposium on Circuits and Systems.* 2016. p. 866–9.

100. Paulsen JA, Renn M, Christenson K, Plourde R. Printing conformal electronics on 3D structures with aerosol jet technology. In: *2012 Future of Instrumentation International Workshop (FIIW) Proceedings.* 2012. p. 1–4.

101. Goth C, Putzo S, Franke J. Aerosol Jet printing on rapid prototyping materials for fine pitch electronic applications. In: *2011 IEEE 61st Electronic Components and Technology Conference (ECTC).* 2011. p. 1211–6.

102. Turner BN., Strong R, Gold SA. A review of melt extrusion additive manufacturing processes: I. Process design and modeling. *Rapid Prototyp J.* 2014;20(3):192–204.

103. Li M, Gou H, Al-Ogaidi I, Wu N. Nanostructured sensors for detection of heavy metals: a review. *ACS Sustain Chem Eng.* 2013;1(7):713–23.

104. Prakash S, Pinti M, Bhushan B. Review article: theory, fabrication and applications of microfluidic and nanofluidic biosensors. *Philos Trans R Soc A Math Phys Eng Sci.* 2012;370(1967):2269–303.

105. Chin CD, Linder V, Sia SK. Lab-on-a-chip devices for global health: past studies and future opportunities. *Lab Chip.* 2007;7(1):41–57.

106. Tsopela A, Laborde A, Salvagnac L, Ventalon V, Bedel-Pereira E, Séguy I, et al. Development of a lab-on-chip electrochemical biosensor for water quality analysis based on microalgal photosynthesis. *Biosens Bioelectron.* 2016;79:568–73.

107. Neethirajan S, Ragavan KV, Weng X. Agro-defense: biosensors for food from healthy crops and animals. *Trends Food Sci Technol.* 2018;73:25–44.

108. Justino CIL, Duarte AC, Rocha-Santos TAP. Recent progress in biosensors for environmental monitoring: a review. *Sensors.* 2017;17(12):2918.

7 Direct Write Printed Electronics and Materials Synthesis Using Non-equilibrium Plasma-Based Techniques

Matthew Montgomery
Yale University
Plazmod Technologies Inc.

Chris Funke
Plazmod Technologies Inc.

Neal Magdefrau
Electron Microscopy Innovative Technologies LLC

Paul Sheedy
Raytheon Technologies Center

Mary Herndon
Raytheon UMass Lowell Research Institute

Wayde Schmidt
Mat-Chem LLC

Andrew S. Morgan
Yale University

Evyatar Shaulsky
Yale University

Shomeek Mukhopadhyay
Yale University

DOI: 10.1201/9781003138945-7

7.1 INTRODUCTION

Additive manufacturing (AM) or direct write printing technologies have the potential to disrupt the world of conventional manufacturing in electronics, in particular for applications requiring flexible and unconventional substrates (curved substrates), while reducing the time required for prototyping and materials cost [1–3]. The substantial reduction of the tooling costs, time to production as well as the ability to incorporate complex multi-material designs make direct write printing techniques extremely attractive for internet of things (IoT), integrated sensors and radiofrequency (RF) applications [3–5]. Essentially, all of the direct write printing technologies have the common goal of being able to print multi-functional microsystems in the length scales ranging from a few microns to millimeters in areas and applications where conventional semiconductor and PCB fabrication technologies are either too expensive or cannot work, e.g. direct printing of biomaterials and tissues. The deposition techniques range from photopolymerization, material extrusion, to direct energy deposition using a wide variety of materials from powders to functional biomaterials. This chapter will focus on the ability to print electronics and sensors using a direct write method that uses a solid wire as a source, operates under ambient or near ambient conditions, is to a large extent surface agnostic and requires minimum post-processing. A significant amount of work in recent years have focused on the use of conductive inks for direct writing of printed circuits.

Despite impressive gains in conductivity, the lack of resiliency in harsh environments [6] and the fact that each new application requires new process and materials chemistry development [7] make the development of complementary techniques very important. Microplasma [8] and liquid metal droplet-based techniques [9] fall in this category. In the next section, we will focus on different microplasma-based techniques as well a hybrid technique which combines plasma with nanopowder-based deposition [10]. A related hybrid technique pioneered in Milan combines standard fused filament technology and supersonic cluster beams to deposit the active, tailored electronic components [11]. One particular direct write technology which requires activation of an active radical containing molded substrate either by light or some other suitable initiator is essentially a hybrid method and is the only method known to the authors, which is used in full scale production for consumer automotive applications, where the challenges of harsh and high-stress (g-forces) environments do not arise [12]. We shall not discuss direct deposition of liquid metal, solder or electrohydrodynamic spinning based techniques since they do not fall under the purview of plasma-based processes.

In 1999, the Defense Advanced Research Projects Agency (DARPA) launched the Mesoscale Integrated Conformal Electronics (MICE) program. The specific aims of this program were to develop technologies that could provide industry with rapid prototyping and just-in-time electronics manufacturing on conformal surfaces. MICE also considered the morphological requirements of the structural surface, where neither the patterning nor the post-processing steps should heat the substrate/surface above its damage threshold. This has led to continued material development of thick film inks with lower processing temperatures. However, plasma-based deposition processes were not part of the original MICE program, and hence, they represent an exciting new frontier for advanced printing and synthesis technologies.

7.2   DIRECT WRITE PLASMA-BASED TECHNIQUES

Plasma processes have been key drivers in the semiconductor industry, and in particular, low-temperature plasmas have been used for decades as a reactive environment for advanced materials fabrication and for functional devices. Common examples of materials and processes that benefit from plasmas include deposition, photoresist etching, surface cleaning for the microelectronic, automotive and several other industries. Low-pressure, low-temperature plasmas have been highly successful in the manufacturing of microelectronic components. Since plasmas are composed of electrons, positive ions and neutral species, they can be classified based on the degree of ionization, energy density and whether the electrons and the heavier gaseous species are in thermodynamic equilibrium. High-density plasmas typically used in etching or producing nanomaterials have particle densities $N > 10^{15-18} \text{cm}^{-3}$. The high number of ions enhances excitation and ionization through collisions and also increases the ion bombardment rate. Plasmas with $N < 10^{12-14} \text{cm}^{-3}$ are considered low density plasmas, e.g. the ones used in laser wakefield accelerators [13]. When the electron temperature T_e, ion temperature T_i and neutral temperature T_n are identical, the plasma is in thermodynamic equilibrium since all the three species follow the same Maxwell Boltzmann distribution. In the case of nonthermal and non-equilibrium plasmas, the electron temperature is substantially higher than the ion and neutral temperatures, $T_e \gg T_i, T_n$, when the momentum transfer between the lighter electrons and the heavier particles is not very efficient and tends to favor electrons. They can be generated by corona, glow or arc discharge and in both inductive and capacitive coupling modes. Atmospheric pressure plasmas are a special class of plasmas produced under ambient pressure or close to ambient pressure conditions. It is relatively easy to produce non-equilibrium gas discharges at low pressures and gas temperatures close to room temperature, e.g. in a fluorescent tube or magnetron system. However, as the pressure is raised, gas discharges become unstable by a glow-to-arc transition mode. Hence, to have a stable glow discharge mode operation under high pressure (atmospheric pressure), one needs to use special electrodes, geometries and/or excitation schemes to maintain a lower gas temperature. Microplasmas are plasmas where the typical dimensions, e.g. between electrodes, are of the order of a few microns to a few millimeters. The typical cathode sheath at atmospheric pressure is ~100 micrometers which means that these plasmas can operate under atmospheric pressure conditions but will be dominated by boundary conditions. A typical rule of the thumb is that the pressure multiplied by the electrode gap should be a constant as exemplified by the Paschen relationship. Since Magnetron glow discharges operating at ~1 Torr have typical electrode-substrate separation of ~10 cm, glow discharges operating at 760 Torr will have typical electrode separation of 100–200 microns. For normal glow discharge, the cathode fall length, d_n (centimeters), varies inversely with the pressure, p (torr) as:

$$pd_n = \ln\left(1 + \frac{1}{\gamma}\right) / \left(\frac{\alpha}{p}\right)$$

The first Townsend coefficient, α, is defined as the number of ionizing collisions per unit path length and γ is the number of electrons emitted per incoming ion, and these constants have to be determined empirically for each gas and electrode combination. Hydrogen gas with iron electrodes at $p = 760$ Torrs gives $d_n = 10$ μm. Once the Townsend coefficients are known, a given operating pressure can guide us to the electrode separation for stable glow discharge operation. Microplasmas have a wide range of applications including materials synthesis, surface disinfection, water treatment, and UV light sources, among others [14,15].

The existence of a stable glow discharge plasma under atmospheric pressure conditions is *necessary but not sufficient* to ensure material deposition. In a plasma, the heavier ions and neutrals can hit the surface of a 'target', which is typically kept at a negative potential. When an ion hits the surface of the target, (1) the ion may be reflected and possibly neutralized, (2) the impact can eject secondary electrons, (3) the ion can be buried or implanted in the target, (4) the ion impact can cause structural changes like vacancies, interstitials and lattice defects and (5) the impact causes a cascading series of collisions between atoms of the target leading to the ejection of one or more target atoms in a process known as 'sputtering'. This process forms the *backbone* of the direct write plasma-based processes. Under a given set of conditions, e.g. pressure, electrode separation, and operating voltage, one or more of the above-mentioned processes will dominate. The 'art' consists in choosing the correct set of conditions so that sputtering will dominate over the other competing processes. It is important to remember that sputtering consists in removing or etching the source or target material and simultaneous deposition of the material on the substrate. If the substrate is not biased properly or if the ejected material is not guided to the correct location, then one cannot use this for direct write processes.

One of the first groups to attempt direct write-based deposition process under ambient conditions (hence microplasma based) was that of the Sankaran group at Case Western Reserve University (currently based at University of Illinois Urbana-Champagne (UIUC)) [16,17]. The system consisted of a thin wire (10–100 micron) which was the source or target and was negatively biased, and the copper disk electrode acted as the anode and the gas flow was coaxial to the target wire – typically argon or nitrogen is used. The wire is placed ~100 micrometers above the substrate and the substrate is moved in the horizontal plane in order to achieve the required pattern. A magnet is placed behind the disk anode to increase the ionization by electrons similar to magnetron sputtering which also stabilizes the plasma locally between the wire and the anode. The entire process uses a DC source with a suitable ballast resistor, and the total process footprint is that of a small desktop computer which is ideal for field applications. The process throughput is essentially low since sputtering is a low throughput process, and since the target geometry is reduced from that of a disk (standard magnetron sputtering) to that of a wire, this reduces the throughput by another order of magnitude. One variation of the technique uses the microplasma as a reducing beam in order to reduce metal salt impregnated surface in the desired pattern [18]. This technique is very similar to the 'plastronique' [12] hybrid method discussed earlier; the main difference is that the reduction with the microplasma happens in the liquid phase. Interestingly, both

Sankaran group and other groups have explored the use of microplasmas for etching in addition to deposition applications [19,20].

Another interesting variation of the wire source-based approach is to use multiple anodes instead of a single disk anode and suitable biasing of the anodes and cathodes in order to have a focused beam of the sputtered material [8]. There are no magnets used in this method and ~100 nm feature sizes with gold were produced.

A significantly different method that uses the barebones sputtering approach without any magnets or focusing electrodes was used [21,22] by generating a highly confined microglow plasma between the target's tip and the substrate at atmospheric pressure. Using cylindrical copper electrodes with diameters ~40 μm, a microplasma is used to deposit material on doped silicon with thicknesses ranging from the 100 nm order to over 5 μm. The process has extremely slow write speeds and can only work with conducting or semiconducting substrates.

One interesting variation of the process which uses a spark plasma to generate the material in the form of nanoparticles and then guide and focus the particles to the desired location has been used [23–26]. This technique is a natural evolution of the direct electric discharge machining (EDM) technique and is significantly faster in terms of material throughput, but since the deposited material is aerosolized, it suffers from the same limitations as other aerosol-based direct write techniques. Another interesting variation of the EDM-based technique in order to increase the throughput is to use patterned masks that are biased through the electrically conducting substrate and the charged particles are electrostatically focused to the open areas of the mask when the entire area is filled from the nanoparticle source [27].

7.3 BACKGROUND OF THE TECHNOLOGY

Majority of the sources for non-equilibrium plasma discharges produce a plasma that is geometrically confined between the electrodes that are, for example, used in dielectric barrier discharge systems. The other ones are the atmospheric pressure plasma jets, which are not confined to the geometric boundaries and are typically used with a high gas flow rate. The plasma (typically the afterglow region) exits the source, and the carrier gas along with the active species impinges on the substrate. In the Plazmod system, the main source is a wire or rod, although for some applications we have used a disk source with suitable modifications [*US Patent Number 10357920 (2019)*]. The plasma creation system is kept well separated from the downstream section, and the external chamber pressure can be varied to create supersonic and even hypersonic conditions. The wire is biased using a high voltage DC (1–4 kV) or a pulsed DC with a short duty cycle running at 33 kHz and 20–30 kV depending on the material. The high velocity carrier gas goes through a section where the source atoms in the plasma chamber are entrained in the carrier gas. The carrier gas moves from a high-pressure zone to a low-pressure zone, which is typically maintained from 100 millitorrs to atmospheric pressure. Note that the flow conditions create hypersonic gas velocities in the downstream section and extremely high kinetic energies; it also can create a sharp focused region of velocity where the substrate can be placed and where one can envision a colliding beam such as non-equilibrium synthesis-like conditions. The four experimental control parameters are chamber pressure, gas identity, gas flow rate and the energy injected in the plasma zone. Figure 7.1 shows a cross section of the apparatus as well as the image of the machine in operation. The rightmost column of images (1) shows the SEM images of the wire source after the deposition in stable glow discharge mode where the fracturing of the wire happened during removal from the machine but the wire face shows erosion patterns similar to magnetron sputtering, (2) shows a SEM image after the wire underwent significant glow to arc transition and (3) is a higher magnification image of the same wire source. In (2), the post-deposition state of the wire shows significant globular-like structure reminiscent of cathodic arc processes where localized melting and redeposition along with the formation of macros occur. Since arc discharges carry significant currents as opposed to glow discharges, random localized melting is well known and one mitigates this by using arc suppression feedback loops and very short (~10 microsecond) duty cycles.

In Figure 7.2, we show some representative simulations of copper atoms in the Argon gas flow. The chamber pressure downstream is set at 100 milli-Torrs and the copper atoms are introduced in the very thin capillary-like structure in the middle of the tube (Figure 7.2, right panel). This simple simulation does not model the details of the plasma, but the mass flow rate is known which goes as an initial condition along with flow rate, radius and temperature of copper atoms, initial temperature of the carrier gas (293K) and the pressure of the outer chamber and the geometry dictated by our experimental conditions. The first three plots are of the gas flow only and the last one is of the copper atoms in the carrier gas. The first

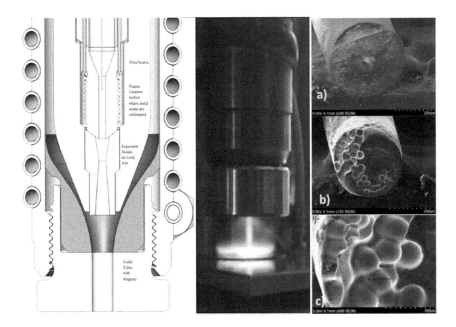

FIGURE 7.1   The schematic of the Plazmod machine (left) and the machine in operation (middle). The rightmost panel shows the state of the wire under different deposition conditions (see the text for details).

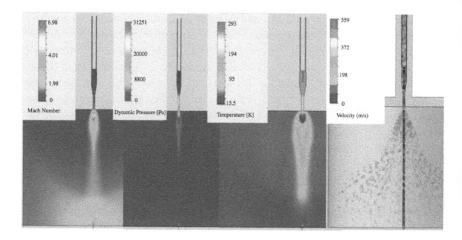

FIGURE 7.2 Representative simulation of the gas flow showing from left to right: (a) Mach number, (b) dynamic pressure, (c) temperature and (d) entrainment of the copper atoms in the gas flow.

plot of the Mach number shows that under realistic experimental conditions, we easily approach hypersonic (>5) conditions, which are consistent with the dynamic pressure and velocity plots. The temperature goes down to around 15K, which is not surprising since the velocity is very high, and the last image is that of entrained copper particles in the Argon flow.

Future work will involve the correlation of the numerical model of the plasma including the afterglow region, which will have to include the detailed reaction mechanism and the addition of Poisson's equation to account for the electron and ionic motion. Known rates of the ion-neutral and electron-neutral collisions will have to be included.

In the next section, we will look at the synthesis of two specific metals, e.g. copper and zinc, both with respect to bulk properties as well as printed structures, and then discuss the synthesis of more complex structures.

7.4   DEPOSITION OF BULK METALS

7.4.1   Copper

Copper is typically used for both electrical connectivity as well as antenna structures since it has high thermal and electrical conductivity which allows the fabrication of complex patterns without losing performance. Initially, silver inks were used for printing by inkjet technology. However, the very high cost of silver limits wide industrial application. Since copper is not only much cheaper but possesses a very high conductivity (only 6% less than that of Ag), copper can be considered as a replacement for silver. Hence, the synthesis of Cu nanoparticles is of great interest from a scientific as well as an industrial point of view, due to its huge potential for replacing the expensive silver ink. In terms of direct deposition of Copper-based nanoparticles (as opposed to inks with binders), previous work has focused on vacuum-based synthesis [28] or wet chemistry based precursor techniques [29]. In the following, we will look at the bulk properties of copper traces deposited on glass substrates.

Figure 7.3 shows white light interferometry of copper deposited on borosilicate glass with Argon gas flow as optimized for our purposes. The lines are typically 5–10 mm in length with a width of ~200 microns and a height of 550 nanometers with a cross section shown in the inset. Even though the vertical walls show no bleeding, sometimes contamination can adhere to the sides which may require an extra cleaning step with acetone to remove the adhering particles. The roughness Ra ~20 nm. The right panel of Figure 7.3 shows the EDS results from the eSEM after removing a suitable section for analysis.

Further morphological analysis of the printed copper traces with the eSEM can be done after sectioning with the FIB as shown in the top panel of Figure 7.4. The evolution of the DC conductivity is shown as a function of different runs. For the case of optimum results in conductivity, we should expect that the print thickness should be 3–4 times the substrate roughness which in the case of the glass substrate translates to about 5 microns. This trend is seen in the prints on Alumina substrate where conductivity is about 30%–40% lower for traces of the same thickness. The inset to the top right figure shows a long rectangular structure which was printed on Silicon Carbide as a testbed for coplanar waveguide structures and to measure insertion losses.

The conductivity of the samples is well within the range needed for printed electronics, but can only be used for some microwave patch antenna applications. In particular, the surface roughness of the copper traces, are typically between 100 and 200 nm depending on the speed of the process. Increasing surface roughness leads to

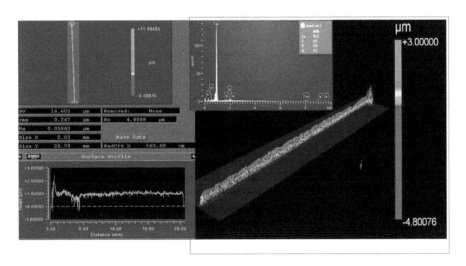

FIGURE 7.3 White Light Interferometry of a copper trace on borosilicate glass (left panel) along with the cross section (below). The right panel shows a rougher deposition with the EDS of a section of the trace. The inset in the top right image shows a part of a coplanar waveguide like structure printed on Silicon Carbide.

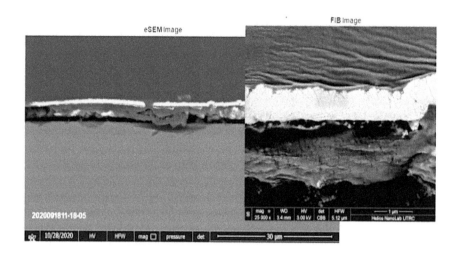

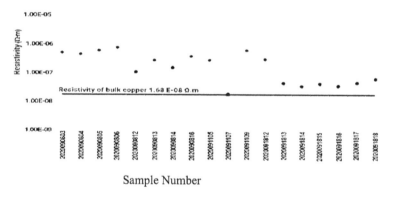

Sample Number

FIGURE 7.4 SEM cross sections of the copper depositions on glass. The surface was pretreated with a hydrogen plasma before copper deposition. The lower panel shows the variation of DC conductivity with optimization of the deposition parameters.

increasing scattering and losses in the coplanar waveguide type antennas; in particular, when the roughness is of the order of skin depth at a given frequency, the transmission losses become significant. With 200 nm roughness, operating at frequencies up to 28.5 GHz should be possible [30].

We next turn to the fabrication of a complex 'strain gauge' type structure to look at the variability in both morphology as well as the measured conductivity to access the suitability of the process for direct printing of circuits. Figure 7.5 (top left panel) shows the details of the printed structure of copper on Alumina (which is an excellent dielectric) where each of the individual traces is marked. The DC conductivity of the traces is shown below for each of the marked traces. With the exception of trace number 1, the others vary by ~12% from the mean. The white light interferometry and the cross-sectional images show that the high-speed nature of this deposition process on a rough surface like alumina tends to create more ripples rather than a flat top like glass or silicon substrate.

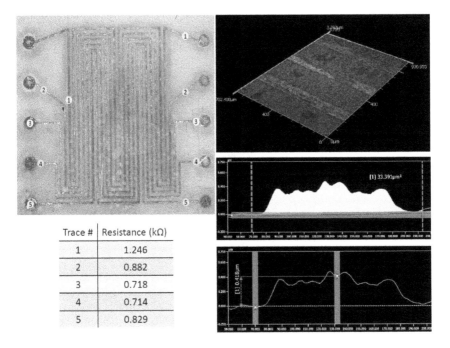

Trace #	Resistance (kΩ)
1	1.246
2	0.882
3	0.718
4	0.714
5	0.829

FIGURE 7.5 Copper traces winding on an Alumina substrate with the electrical measurement of each trace (left panel). The right panel shows a typical white light image along with a cut of the cross section.

These structures may be quite suitable for applications in conformal electronics on challenging substrates like alumina or silicon carbide, but they will not be suitable for applications in RF or microwave antennas.

7.4.2 ZINC

Zinc is a very useful metal used in entirely different industrial settings as compared to, for example, Copper or Aluminum. In particular, coatings represent the largest single use for zinc. This particular use has increased significantly as a result of the necessity to conserve steel and make the most efficient use of materials and energy. In addition to energy savings applications, it is a good material to investigate the nature of crystal growth in the deposition process as ultrathin films were deposited on smooth silicon wafers.

The zinc foils shown in Figure 7.6 were deposited on polished silicon wafers using 100 micron zinc wires as the source material. These foils were typically 1 micron in thickness and surface roughness of less than 50 nanometers by profilometry, which is significantly less than that for copper depositions. Interestingly, the grain boundary on the side interfaced with the silicon wafer appears shiny visually and has smaller grain sizes (SEM images on the top panel) compared to the side exposed to air which appears significantly dull in appearance even though the grain sizes are larger. It was thought that there was an oxide layer on the exposed surface which turned out not

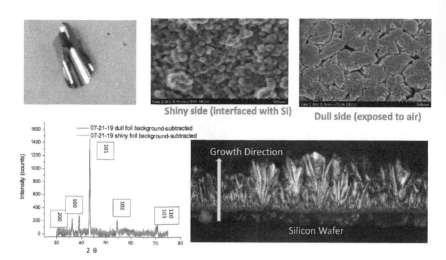

FIGURE 7.6 Zinc deposited on silicon wafers (top left) is visually shiny on the wafer side, but the grain structure is very different depending on the side facing the wafer or exposed (top panel). The X-ray diffraction data with the Miller indices show that the crystal structure of both sides is identical and equivalent to bulk crystals. The metallographic image on the bottom right shows columnar growth without any inclusions or pores.

to be the case after performing thin film XRD. The crystal structure of a typical metal produced is indistinguishable from that produced by other means. The Miller indices are an exact match with those in literature, but more importantly, there are no contaminating peaks, e.g. ZnO and both sides (shiny and dull) have exactly the same peaks [31]. Metallurgical analysis under microscopy after cutting and polishing the metal shows the columnar growth of grains in the direction away from the silicon surface toward the source. This fast deposition technique produces higher quality films than electrodeposition and is of great interest in zinc air batteries and high-current Zinc/AgO high-current bipolar batteries [32]. The higher quality of the films is expected to suppress dendrite formation which is extremely important in the development of solid-state electrolyte systems. Our future work plans include extending this to pure lithium anode for battery applications. In fact, there have been suggestions that zinc ion batteries can be a scalable alternative to lithium ion batteries given extensive supply chain and refining problems with lithium, at least for certain applications. In particular, recent work has shown that pairing zinc with nickel can be a viable alternative to a lithium based battery [33].

In order to deposit more complex materials beyond binary transition metal oxides [34–37], we used a disk source with an argon blanket and extremely low gas flow rates which are typically around 100 sccm. Significant care must be taken to keep the power supply in the glow discharge mode and prevent a glow to arc transition which can cause runaway surface melting processes as discussed in Figure 7.1. We chose to look at Lithium Lanthanum Zirconium Oxide or LLZO, which is an important ionic conductor for Lithium in its cubic/garnet phase [38,39]. Since this is one of the candidate electrolytes for solid-state batteries, it is currently synthesized using standard

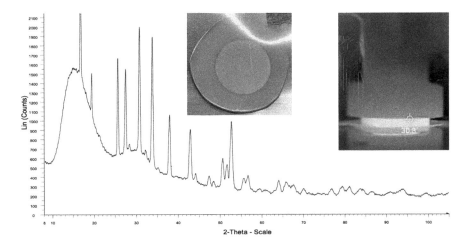

FIGURE 7.7 The synthesis of LLZO from a disk source (top right inset). Infrared image shows the low-temperature aspect of the deposition. The LLZO is printed directly on the copper current collector (top middle inset). The XRD of the LLZO shows the correct garnet phase along with a dispersed amorphous phase.

ceramic manufacturing techniques, which involve sintering to remove binders. The main challenge for an atmospheric pressure plasma-based technique was to get to the target structure of the correct size on a copper substrate which traditionally acts as a current collector. In Figure 7.7, we show the XRD of the target LLZO garnet structure which shows all the main peaks. However, an amorphous background shows that some non-crystalline phases are possibly present in the crystal phase.

7.5 OTHER COMPLEX STRUCTURES

One of the significant applications for direct printed structures is monitoring strain and temperature in harsh and challenging environments while reducing both weight and drag. One of the areas where this is of particular concern is the size and number of current antenna systems. A typical aircraft can have anywhere from 20 to 50 different antennas protruding from its surface. These antennas can cause considerable drag, increased fuel consumption and additional weight. Defense systems also have requirements for antennas that have covert or stealthy properties such that the antenna does not backscatter radiation when illuminated by a hostile transmitter. For sensing requirements, printed strain gauges can be used to detect blade rub in jet engines and printed resistance temperature detectors (RTDs) for 2D temperature mapping. For some requirements, antennas can first be printed on a stretchable network, then applied to a substrate. In Figure 7.8, we show some examples of these printed structures which include strain gauges, spiral antenna which can act as broadband detectors and a recirculating waveguide structure which is directly printed on an alumina substrate.

Some of the other applications which we have explored but do not go into detail in this chapter are direct and maskless etching using mixtures of SF_6 and other suitable

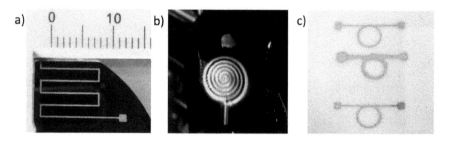

FIGURE 7.8 Strain gauge structure of copper on silicon carbide (a), spiral structure of titanium on silica (b) and recirculating waveguide structures of copper on 15 mm alumina substrates (c).

gases [40], printing of filters for THz applications and cleaning contaminants like PFAS (also known as forever chemicals) in water [41].

7.6 CONCLUSION

Direct write technologies have significantly evolved over the past decades since the initiation of DARPA's original MICE program. Ink and jet-based technologies are now routinely used in prototyping and limited production runs. However, directly printed RF structures, especially for harsh environments in defense applications, still have a significant path left for maturation before they can be deployed. In particular, other technologies need to be explored and one of the most promising technologies is based on atmospheric pressure plasma-based depositions. In this chapter, we have looked at some of the material properties as well as the suitability of the technique for deposition of complex structures. Another significant advantage of the plasma-based techniques is that they allow 'in situ' complex materials synthesis, which has enormous potential in energy applications.

ACKNOWLEDGMENTS

We would like to thank Gilad Kaufman, Mike Leddy, Zachary Fishman, Manoj Kolel-Veetil and Mike Hollenbeck for useful discussions and help with some of the measurements. This work was partly supported by NSF grants 1645623 (2017), 1745845 (2018) and DOE grant 245845 (2019).

REFERENCES

1. Bingheng, L., L. Hongbo, and L. Hongzhong, Additive manufacturing frontier: 3D printing electronics. *Opto-Electronic Advances*, 2018. **1**(1): p. 170004.
2. Gibson, I., et al., *Development of Additive Manufacturing Technology, in Additive Manufacturing Technologies*, I. Gibson, et al., Editors. 2021, Springer International Publishing: Cham. p. 23–51.
3. Sachs, E., M. Cima, and J. Cornie, Three-dimensional printing: rapid tooling and prototypes directly from a CAD model. *CIRP Annals*, 1990. **39**(1): p. 201–204.

4. Flowers, P.F., et al., 3D printing electronic components and circuits with conductive thermoplastic filament. *Additive Manufacturing*, 2017. **18**: p. 156–163.
5. Longtin, J., et al. Sensors for harsh environments by direct write thermal spray. in *SENSORS, 2002 IEEE*. 2002.
6. Neff, C., et al., A fundamental study of printed ink resiliency for harsh mechanical and thermal environmental applications. *Additive Manufacturing*, 2018. **20**: p. 156–163.
7. Chang, J., et al., Advanced material strategies for next-generation additive manufacturing. *Materials*, 2018. **11**: p. 166.
8. Kornbluth, Y.S., et al., Room-temperature, atmospheric-pressure microsputtering of dense, electrically conductive, sub-100 nm gold films. *Nanotechnology*, 2019. **30**(28): p. 285602.
9. Murr, L.E. and W.L. Johnson, 3D metal droplet printing development and advanced materials additive manufacturing. *Journal of Materials Research and Technology*, 2017. **6**(1): p. 77–89.
10. Gandhiraman, R.P., et al., Plasma jet printing of electronic materials on flexible and non-conformal objects. *ACS Applied Materials & Interfaces*, 2014. **6**(23): p. 20860–20867.
11. Bellacicca, A., T. Santaniello, and P. Milani, Embedding electronics in 3D printed structures by combining fused filament fabrication and supersonic cluster beam deposition. *Additive Manufacturing*, 2018. **24**: p. 60–66.
12. Cheval, K., et al., Progress in the manufacturing of molded interconnected devices by 3D microcontact printing. *Advanced Materials Research*, 2014. **1038**: p. 57–60.
13. Liang, E., Comoving acceleration of overdense electron-positron plasma by colliding ultra-intense laser pulses. *Physics of Plasmas*, 2006. **13**(6): p. 064506.
14. Chiang, W.-H., et al., Microplasmas for advanced materials and devices. *Advanced Materials*, 2020. **32**(18): p. 1905508.
15. Mariotti, D. and R.M. Sankaran, Microplasmas for nanomaterials synthesis. *Journal of Physics D: Applied Physics*, 2010. **43**(32): p. 323001.
16. Sankaran, R.M. and K.P. Giapis, High-pressure micro-discharges in etching and deposition applications. *Journal of Physics D: Applied Physics*, 2003. **36**(23): p. 2914–2921.
17. Chiang, W.-H., C. Richmonds, and R.M. Sankaran, Continuous-flow, atmospheric-pressure microplasmas: a versatile source for metal nanoparticle synthesis in the gas or liquid phase. *Plasma Sources Science and Technology*, 2010. **19**(3): p. 034011.
18. Richmonds, C. and R.M. Sankaran, Plasma-liquid electrochemistry: Rapid synthesis of colloidal metal nanoparticles by microplasma reduction of aqueous cations. *Applied Physics Letters*, 2008. **93**(13): p. 131501.
19. Sankaran, R.M. and K.P. Giapis, Maskless etching of silicon using patterned microdischarges. *Applied Physics Letters*, 2001. **79**(5): p. 593–595.
20. Ichiki, T., R. Taura, and Y. Horiike, Localized and ultrahigh-rate etching of silicon wafers using atmospheric-pressure microplasma jets. *Journal of Applied Physics*, 2004. **95**(1): p. 35–39.
21. Abdul-Wahed, A.M., et al., Direct writing of thin and thick metal films via micro glow plasma scanning, in *2016 IEEE 29th International Conference on Micro Electro Mechanical Systems (MEMS)*. 2016.
22. Abdul-Wahed, A.M., A.L. Roy, and K. Takahata, Microplasma drawing of thermocouple sensors, in *2016 IEEE Sensors*. 2016.
23. Jayasinghe, S.N., M.J. Edirisinghe, and D.Z. Wang, Controlled deposition of nanoparticle clusters by electrohydrodynamic atomization. *Nanotechnology*, 2004. **15**(11): p. 1519–1523.
24. Meuller, B.O., et al., Review of spark discharge generators for production of nanoparticle aerosols. *Aerosol Science and Technology*, 2012. **46**(11): p. 1256–1270.
25. Kim, H., et al., Parallel patterning of nanoparticles via electrodynamic focusing of charged aerosols. *Nature Nanotechnology*, 2006. **1**(2): p. 117–121.

26. Cole, J.J., et al., Continuous nanoparticle generation and assembly by atmospheric pressure arc discharge. *Applied Physics Letters*, 2009. **95**(11): p. 113101.
27. Fang, J., et al., Approaching gas phase electrodeposition: process and optimization to enable the self-aligned growth of 3D nanobridge-based interconnects. *Advanced Materials*, 2016. **28**(9): p. 1770–1779.
28. Liu, Z. and Y. Bando, A novel method for preparing copper nanorods and nanowires. *Advanced Materials*, 2003. **15**(4): p. 303–305.
29. Huang, H.H., et al., Synthesis, characterization, and nonlinear optical properties of copper nanoparticles. *Langmuir*, 1997. **13**(2): p. 172–175.
30. Huang, B. and Q. Jia, Accurate modeling of conductor rough surfaces in waveguide devices. *Electronics*, 2019. **8**(3): p. 269.
31. Chaba, N., et al., Morphology study of zinc anode prepared by electroplating method for rechargeable Zn-MnO(2) battery. *Heliyon*, 2019. **5**(10): p. e02681.
32. Galvelis, R., et al., Comparison of the relative stability of zinc and lithium-boron zeolitic imidazolate frameworks. *CrystEngComm*, 2012. **14**(2): p. 374–378.
33. Parker, J.F., et al., Rechargeable nickel-3D zinc batteries: an energy-dense, safer alternative to lithium-ion. *Science*, 2017. **356**(6336): p. 415–418.
34. Shuiabov, A., et al., Synthesis of nanostructured transition metal oxides by a nanosecond discharge in air with assistance of the deposition process by plasma UV-radiation. *Advances in Natural Sciences: Nanoscience and Nanotechnology*, 2018. **9**(3): p. 035016.
35. Sun, D., et al., Atmospheric pressure plasma-synthesized gold nanoparticle/carbon nanotube hybrids for photothermal conversion. *Langmuir*, 2019. **35**(13): p. 4577–4588.
36. Marino, E., et al., Synthesis and coating of copper oxide nanoparticles using atmospheric pressure plasmas. *Surface and Coatings Technology*, 2007. **201**(22): p. 9205–9208.
37. Mariotti, D., et al., Low-temperature atmospheric pressure plasma processes for "green" third generation photovoltaics. *Plasma Processes and Polymers*, 2016. **13**(1): p. 70–90.
38. Kravchyk, K.V., D.T. Karabay, and M.V. Kovalenko, On the feasibility of all-solid-state batteries with LLZO as a single electrolyte. *Scientific Reports*, 2022. **12**: p. 1177.
39. Sastre, J., et al., Blocking lithium dendrite growth in solid-state batteries with an ultrathin amorphous Li-La-Zr-O solid electrolyte. *Communications Materials*, 2021. **2**: p. 76.
40. Jeong, J.Y., et al., Etching materials with an atmospheric-pressure plasma jet. *Plasma Sources Science and Technology*, 1998. **7**(3): p. 282–285.
41. Lewis, A.J., et al., Rapid degradation of PFAS in aqueous solutions by reverse vortex flow gliding arc plasma. *Environmental Science: Water Research & Technology*, 2020. **6**(4): p. 1044–1057.

8 Additively Manufactured Ceramics with Embedded Conductors for High-Temperature Applications

Eleanor Rogenski
Youngstown State University

Victoria Admas
Youngstown State University

Bhargavi Mummareddy
Youngstown State University

Bradley Duncan
Lincoln Laboratory Massachusetts Institute of Technology

Eric MacDonald
The University of Texas at El Paso

Pedro Cortes
Youngstown State University

Compared to the conventional materials used in the additive manufacturing field, ceramics offer advantageous properties at high temperatures, such as hardness, strength, rigidity, wear, and corrosion resistance, which are required in a wide range of aerospace applications [1–4]. Similarly, ceramic systems offer viable thermal, optical, electrical, and magnetic features that have made them attractive materials in the electronics field [5]. On the other hand, ceramics also present a number of limitations including ductility, shock resistance, and dimensional tolerance, which has constrained their ample production across different additive manufacturing technologies when compared to polymers and metals [6,7]. The 3D printing process of ceramic materials remains a challenging procedure due to difficulty in creating

DOI: 10.1201/9781003138945-8

dense parts while avoiding cracking and delamination. The additive technologies for producing ceramics can be categorized by feedstock type: slurry, powder, and bulk solid. Table 8.1 summarizes different printing technologies used for producing ceramic parts.

The type of manufacturing process selected strongly affects the resolution, quality, and performance of the printed part. Indeed, the level of porosity, cracking, thermal stresses, and surface finish depend on the manufacturing process. Additionally, most ceramic processing technologies require a thermal post-treatment, which impacts the final material properties, most notably density and strength. Current efforts on printed ceramics have shown that the slurry-based feedstock produces the parts with the highest densities. The current NanoParticle Jetting™ (NPJ) technology used by the XJET Carmel 1400 3D printer involves the dispersion of ceramic nanoparticles, suspended in a liquid solution. This technology produces highly detailed zirconia parts with a layer thickness of 10 microns and a resolution of 20 microns (Figure 8.1).

TABLE 8.1
Summary of Different Additive Manufacturing Systems to Fabricate Ceramic Parts [5]

Feedstock Form	AM Technology	3D Printing Process
Slurry-based	Vat photopolymerization	SLA (Stereolithography)
		DLP (Digital Light Processing)
	Material jetting	NPJ (NanoParticle Jetting)
Bulk-solid based	Material extrusion	FFF (Fused filament fabrication)
	Sheet lamination	LOM (Laminated Object Manufacturing)
Powder-based	Direct energy deposition	LENS (Laser Engineered Net Shaping)
	Powder-based fusion	SLS (Selective Laser Melting)
		SLM (Selective Laser Sintering)
	Binder jetting	BJP (Binder Jetting Printing)

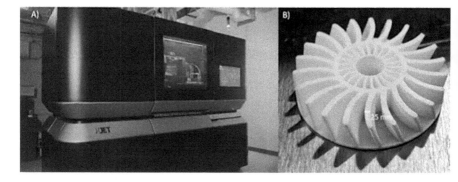

FIGURE 8.1   XJET printing unit. The system uses a NanoParticle Jetting technology capable of producing highly densified parts.

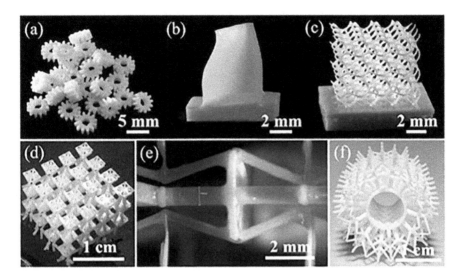

FIGURE 8.2   A variety of sintered ceramic parts produced via Vat Polymerization (i.e., gears and lattice structures) [9,10].

Efforts on vat photopolymerization (VPP) have also resulted in high-resolution ceramic parts (see Figure 8.2). The production of ceramics using VPP can be challenging since a high volume fraction of ceramic particles in the resin is commonly needed for reduced shrinkage as well as to produce a high density part after sintering. However, low loading volume fraction is usually required to decrease the viscosity, enhance a homogeneous suspension, and reduce the light scattering effect during the printing process, which can affect the dimensional accuracy [8].

Additional printing processes such as binder jetting, material extrusion, and selective laser sintering have been used to create ceramic structures; however, the porosity left in the material can compromise the mechanical properties as well as decrease the dielectric performance [11]. Similarly, the surface finish left by these processes is significantly rough which often results in poor metallic interconnections when used as microelectronic substrates [12–14]. To overcome these drawbacks, parts created under these technologies usually require post-infiltration processes to enhance the density, reduce the porosity, and improve the surface finishing [15–18].

Several ceramic materials such as silica, silicon carbide, mullite, silicon nitride, and hydroxyapatite have been successfully printed using different additive manufacturing [19–22]. However, among different available ceramics, zirconia, alumina, and barium titanate have attracted considerable attention due to their electromagnetic behaviors such as high relative permittivity, dielectric strength, and low loss [13,23–25]. These systems are good candidates to be used as dielectric resonators for miniaturized microwave antennas and filters [24,26,27]. Indeed, zirconia is an attractive material for radiofrequency (RF) applications, providing a high relative permittivity and low loss tangent, and preventing unnecessary radiation absorption

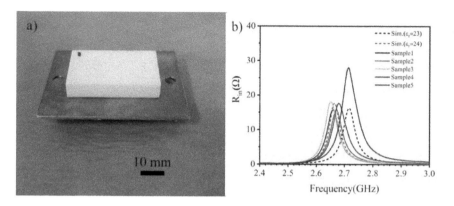

FIGURE 8.3   3D printed ZrO$_2$ electromagnetic test coupon to measure loss and permittivity [28].

FIGURE 8.4   A spatially varying lattice to allow for the modulation of permittivity in 3D space.

and signal attenuation [28]. Figure 8.3 shows a 3D printed zirconia electromagnetic test coupon to measure loss and permittivity, where the parallel resonance was measured between 2.65 and 2.73 GHz.

Alumina has also been used in the electronics area as a packaging material and a substrate for hybrid circuits and multi-layer interconnection circuits. Alumina has a high melting point (2,072°C) with a moderate dielectric constant and a relatively high thermal conductivity (35.4 W/m K) [29]. Figure 8.4 shows a spatially graded lattice ceramic structure manufactured using a Digital Light Processing (DLP) unit (a form of vat photopolymerization using an image projector). The resolution is

30 microns, and this kind of graded system allows density variation throughout the volume to tailor the mechanical, thermal, or electromagnetic performance of the structure.

Current mechanical and hydraulic systems in the aerospace and industrial sectors are being replaced by microelectronic devices in order to increase the performance and reduce weight. These replacements imply that the location of the microelectronic systems is close to engines or under high-temperature conditions. These high-temperature electronic devices require insulation from the heat generation while at the same time removing the heat self-generated by the electrical components. Therefore, the ceramic substrates need to provide a number of thermal features to satisfy these requirements. For instance, the thermal conductivity is one of the critical properties of the ceramic substrates to serve as a heat sink. Similarly, it has been demonstrated that failure of these ceramic platforms is mainly associated with thermomechanical aging; thus, the ceramics should have a similar coefficient of thermal expansion (CTE) to that exhibited by the neighboring electrical parts [30]. The substrate should also have high resistivity to reduce leakage current, a high dielectric strength to withstand voltage, an appropriate dielectric constant to limit the common-mode capacitance, and a relatively good mechanical strength to resist the constraints of handling, vibrations and shocks [31]. Lastly, novel conformal structural designs are required to generate lightweight devices, a requirement that can be satisfied via the complex geometries provided by additive manufacturing.

Nitrate-based ceramics are one of the most attractive systems in the electrical applications; however, their production as 3D printed parts is still under development (particularly for VPP technology) to deliver components with high-resolution features. In contrast, as previously shown, the 3D printing process of alumina and zirconia is robust enough for producing ceramic structures that can address the aforementioned thermal and mechanical requirements. Depending on the packing technology and the needed thermal management, silica can also be considered as an optional system. The thermal and electrical properties of alumina, zirconia, and silica are shown in Table 8.2.

The incorporation of metal traces on complex structures has also been investigated by Oh et al. [28]. The group printed an undulating sinusoidal zirconia part with NanoParticle Jetting, followed by a silver ink deposition process in a double spiral arm using an aerosol jetting system (Figure 8.5). The surface analysis showed

TABLE 8.2
Physical Properties of High-Temperature Ceramics for Electrical Applications [32]

	Thermal Conductivity (W/m K)	CTE (10E-6/K)	Dielectric Strength (kV/mm)
Alumina	26–35	7–9	10–20
Zirconia	1.7–2.7	2.3–12.2	10–23
Silica	1.3–1.5	0.55–0.75	3.4–4.4

 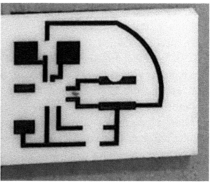

FIGURE 8.5   3D printed NanoParticle Jetting™ ceramic structures to be used as conductive ink platforms: spiral antenna (left) and ATO-based electrical interconnections (right).

a variation of about 5 microns across the surface of the printed zirconia; however, when depositing aerosol jetting connections, the lines are in the submicron range, and consequently, a surface finish improvement in the submicron scale, such as polishing or coating, on the printed ceramic is required to ensure electrical conductivity.

Certainly, while silver can easily support temperatures above 300°C (conditions commonly encountered in small engines), it is not a high-temperature material; it melts about 962°C. Thus, alternative solutions have to be explored in the context of sintering ceramics. One attractive option is the incorporation of high-temperature inks such as Molybdenum- ($T_m = 2{,}623$°C), Tungsten- ($T_m = 3{,}422$°C), or Tungsten-based alloys. Current efforts on aerosol-jetted Molybdenum (Mo) and Antimony Tin Oxide (ATO) on 3D-printed zirconia have shown promising results (Figure 8.5). One drawback of these 3D printed structures is the exposure of the metallic interconnections to either scratches or delamination. An alternative approach to reduce such exposure is the incorporation of channels that can entrap the conductive inks – both protecting from scratching and delamination, but also to mechanically enhance the metal-ceramic adhesion as the structures shrink during the sintering stage. The inclusion of an over-molding approach also represents a promising mechanism to fully encapsulate the metallic phase to eliminate delamination especially during the sintering stage of materials due to the dissimilar CTE.

Current efforts have concentrated on creating designs of diverse ceramic platforms containing different types of conduits or trench cross-sections, i.e., straight-circular, straight-triangular, sinusoidal-circular, etc. for entrapping the metallic connections (Figure 8.6).

The designs have been printed in a Formlabs unit equipped with a 405 nm UV-laser source. The printing process was based on a 50-micron layer thickness, with each printed layer being exposed to 44 seconds of UV light. After the printing stage, the samples were cleaned in isopropyl alcohol (IPA) for approximately 15 minutes to remove excess resin and then heated to 120°C for 75 minutes with a ramp rate of 5°C/min to achieve a curing stage. Here, Formlabs commercial silica resin and Tethon (Nebraska, USA) alumina and zirconia resins have been used in the Formlabs (Massachusetts, USA) unit to produce the parts. The initial attempts of the

FIGURE 8.6   Schematic diagrams of ceramic designs for entrapping metallic depositions.

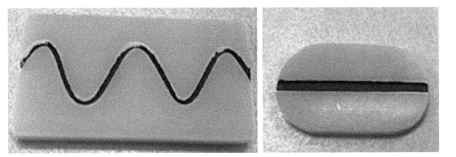

FIGURE 8.7   3D printed cured silica parts with embedded conduits for entrapping conductive inks or pastes. Parts had 20 mm width.

integration of conductive materials on the ceramic platforms have taken place before the sintering stage of the assembled parts (Figure 8.7).

Here, the ceramic part and the metallic phase had different sintering profiles, and to ensure a successful metal conduction and to limit any metal-ceramic delamination, different sintering profiles had to be integrated in the manufactured structures both before and after ink deposition. Especially, when working with two dissimilar systems such as silica (which has an optimal sintering temperature of about 1,275°C) and a Tungsten alloy, which sinters preferably at 1,400°C. Additional studies have shown that considering the initial sintering stage of the ceramic followed by the deposition of the metallic traces and a second sintering process to consolidate the metal phase results in a mechanical robust system without the challenges of metal-ceramic delamination (Figure 8.8).

The sintering profiles of the silica, alumina, and zirconia are shown in Figure 8.9, which have been optimized as a function of the mechanical performance of the manufactured parts. Overall, the parts in the longest direction demonstrate an 18%, 19%, and 22% shrinkage in the silica, alumina, and zirconia parts, respectively.

Several inks and pastes have been investigated as the metallic conductive phase in the 3D printed ceramic platforms. Current efforts have especially concentrated on a high-temperature paste developed by Massachusetts Institute of Technology Lincoln Labs (MIT-LL) based on tungsten-nickel alloys for compatibility in high-temperature applications. Indeed, additional materials such as Mo and ATO from Applied Nanotech, Inc (Austin, Texas, USA) have also been investigated as metallic interconnections. The use of ATO is also of great interest since the ink does not have low

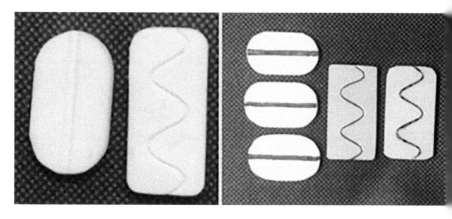

FIGURE 8.8   Sintered ceramic structures (left) followed by the deposition of metallic traces (right) to produce conductive high-temperature platforms. Parts are all 20 mm wide.

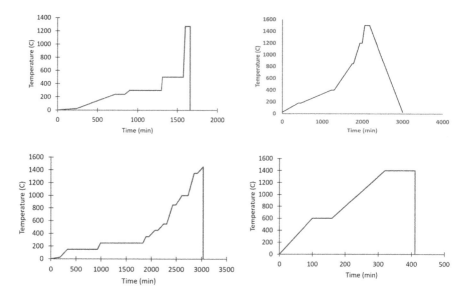

FIGURE 8.9   Sintering profiles of the silica (top left), alumina (top right) and zirconia (bottom left) substrates printed on a VPP unit. Included in the figure is the sintering profile of the W-Ni alloy paste developed by MIT-LL (bottom right) [11,33].

electrical conductivities, which could be an advantageous feature when developing high-temperature electronics. Additionally, the associated sintering process does not require an inert environment. ATO can be sintered in the presence of oxygen, a condition that oxidizes W-Ni and reduces conductivity for electrical interconnections.

Included in Figure 8.9 is the sintering profile of the W-Ni system. Additional parts have also been printed in diverse VPP technologies such as the Tethon Bison unit, the Admaflex-130 as well as open-source SLA systems. Here, continuous studies for optimizing the printing features and sintering outcomes of the printed parts have resulted in parts with superior surface finishing (Figures 8.10 and 8.11).

The conductivity analysis of the sintered parts was evaluated with a multimeter and the measurement shows that the conductive inks sintered sufficiently well to provide a good degree of conductivity with values as low as 0.14 ohm/cm. Further optimizations are in progress to increase the conductivity through optimized sintering of the ceramic metallic structure. Figure 8.12 shows the simple electrical tests performed on the sintered high-temperature printed parts.

FIGURE 8.10   3D printed and sintered alumina structures with sintered ATO (left) and Mo (right) conductive pastes. Parts have 20 mm width.

 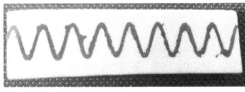

FIGURE 8.11   3D printed sintered zirconia structures. Xjet material with two different materials (Mo and ATO) for producing a high-temperature thermocouple sensor [left]. Inclusion of W-Ni paste in Tethon zirconia. Parts have 20 mm width.

FIGURE 8.12   Conductivity analysis on the sintered parts based on W-Ni paste and zirconia substrates.

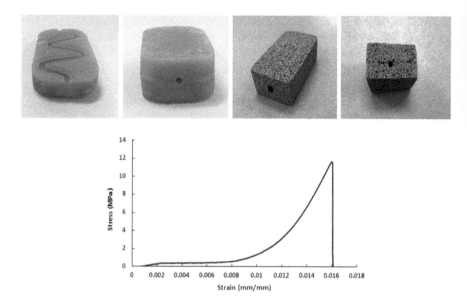

FIGURE 8.13   Over-molded printed parts to encapsulate the metallic traces. Vat photopolymerization printed silica part on the top left and binder jetted alumina hollow spheres on top right. Included in the figure is the averaged flexural strength of the sintered silica over-molded system.

Additional studies were implemented on over-molded ceramic parts in order to fully encapsulate and protect the metallic materials. The over-molding has been investigated in zirconia, alumina, and silica. Briefly, the process consists of joining the "green" printed ceramic parts for the solid-state diffusion to act as the bonding mechanism during the sintering process. This process allows for the post-printing assembly of green parts to increase the size of possible geometries as well as to interpose internal ink traces to serve as high-performance interconnects in arbitrary routes through the ceramic structure – either embedded or on superficial surfaces.

Figure 8.13 shows 3D printed ceramic parts subjected to an over-molding process. A mechanical testing has been executed and has shown similar flexural strength (12 MPa) to that observed on non-over-molded systems, suggesting a high degree of diffusion at the boundary. The over-molding and sintering process has also been performed with embedded W-Ni, where the conductivity of the metal phase has shown a low resistivity (Figure 8.14). Here, hollow alumina spheres were printed in a binder jetting system and used as the encapsulating substrate. The parts had approximately 40% of porosity with a rough surface finishing that would preclude their use with conductive inks. However, a viscous W-Ni paste was used to achieve conductivity values of about 0.09 ohms/cm. These alumina spheres resulted in strong lightweight parts (density = 1.002 g/cm³) with potential applications for the aerospace sector.

The over-molding concept has also been extended into the initial production of low-temperature co-fired ceramics (LTCC). Here, 3D printed alumina structures based on an over-molded block design have been manufactured including

FIGURE 8.14   Over-molded sintered 3D printed parts. Silica/W-Ni structure manufactured using VPP (left). Alumina/W-Ni part based on hollow spheres using binder jetting (right).

vertical interconnections as observed in Figure 8.15. The printed parts displayed in Figure 8.15 are in the green state, and therefore, it shows a high resistivity. This high resistance is highly reduced after the sintering state. Additional efforts are being focused on creating intricate three-dimensional ceramic platforms in which thermal management can be integrated with unprecedented freedom.

8.1   CONCLUSIONS

This work has described the efforts on producing 3D printed ceramic materials such as alumina, silica, and zirconia capable of enduring high-temperature conditions. The work has also shown the incorporation of high-temperature metallic connections such as tungsten-nickel, antimony tin oxide, and molybdenum into the ceramic substrates to yield conductive structures. Following a series of sintering steps, the production of mechanical robust ceramic platforms encasing electrically conductive metallic connections has been demonstrated. An over-molding process has also been studied, and the successful high-temperature embedded systems have been manufactured and tested. The current 3D printing technologies used in this work have allowed the production of functional-hybrid systems that can further be implemented into complex-intricate geometries with unique and customized thermo-managing and conformal designs.

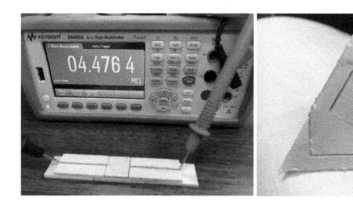

FIGURE 8.15   Green state Alumina-ATO LTCC concept under development (left). Complex printed three-dimensional structures under investigation for integrating electrical and thermal management components.

ACKNOWLEDGMENTS

We would like to thank the Eynon-Beyer Endowment at Youngstown State University and the Murchison Chair at the University of Texas at El Paso for supporting this project. The present work has also been funded by the Assured Digital Microelectronics Education and Training Ecosystem from the AFRL (FA 8650-20-2-1136) and by the High-Temperature 3D-Printed Ceramic Sensor from Marshall NASA grant. DISTRIBUTION STATEMENT A. Approved for public release. Distribution is unlimited. This material is based upon work supported by the Under Secretary of Defense for Research and Engineering under Air Force Contract No. FA8702-15-D-0001. Any opinions, findings, conclusions, or recommendations expressed in this material are those of the author(s) and do not necessarily reflect the views of the Under Secretary of Defense for Research and Engineering.

REFERENCES

[1] Nunomura S, Nakayama J, Abe H, Kamigaito O, Takahara K, Matsusue K. *12 – Mechanical Properties.* In: Sōmiya S, editor. New York: Advanced Technical Ceramics, Academic Press; 1989, pp. 223–58.

[2] Naslain R. Design, preparation and properties of non-oxide CMCs for application in engines and nuclear reactors: an overview. *Compos Sci Technol* 2004;64:155–70.

[3] Rice R. *Mechanical Properties. Cellular Ceramics*, Weinheim, FRG: Wiley-VCH Verlag GmbH & Co. KGaA; 2006, pp. 289–312.

[4] Stambouli AB, Traversa E. Solid oxide fuel cells (SOFCs): a review of an environmentally clean and efficient source of energy. *Renew Sust Energy Rev* 2002;6:433–55.

[5] Chen Z, Li Z, Li J, Liu C, Lao C, Fu Y, et al. 3D printing of ceramics: a review. *J Eur Ceram Soc* 2019;39:661–87.

[6] Hammel EC, Ighodaro OL-R, Okoli OI. Processing and properties of advanced porous ceramics: an application based review. *Ceram Int* 2014;40:15351–70.

[7] Rahaman MN. *Ceramic Processing and Sintering.* New York: CRC Press; 2017.

[8] Gentry SP, Halloran JW. Depth and width of cured lines in photopolymerizable ceramic suspensions. *J Eur Ceram Soc* 2013;33:1981–8.

[9] Schwentenwein M, Homa J. Additive manufacturing of dense alumina ceramics. *Int J Appl Ceram Technol* 2015;12:1–7.

[10] Scheithauer U, Schwarzer E, Moritz T, Michaelis A. Additive manufacturing of ceramic heat exchanger: opportunities and limits of the lithography-based ceramic manufacturing (LCM). *J Mater Eng Perform* 2018;27:14–20.

[11] Bhargavi M, Dylan N, Bharambe VT, Oh Y, Edward B, Ahlfors M, et al. Mechanical properties of material jetted zirconia complex geometries with hot isostatic pressing. *Advances in Industrial and Manufacturing Engineering n.d.*

[12] Karapatis NP, van Griethuysen JPS, Glardon R. Direct rapid tooling: a review of current research. *Rapid Prototyping J* 1998;39:88.

[13] Wilkes J, Hagedorn Y, Meiners W, Wissenbach K. Additive manufacturing of ZrO2–Al2O3 ceramic components by selective laser melting. *Rapid Prototyping J* 2013;39:88.

[14] Agarwala MK, Jamalabad VR, Langrana NA, Safari A, Whalen PJ, Danforth SC. Structural quality of parts processed by fused deposition. *Rapid Prototyping J* 1996;39:88.

[15] Zocca A, Colombo P, Gomes CM, Günster J. Additive manufacturing of ceramics: issues, potentialities, and opportunities. *J Am Ceram Soc* 2015;98:1983–2001.

[16] Nan B, Yin X, Zhang L, Cheng L. Three-dimensional printing of Ti3SiC2-based ceramics. *J Am Ceram Soc* 2011;94:969–72.

[17] Shahzad K, Deckers J, Boury S, Neirinck B, Kruth J-P, Vleugels J. Preparation and indirect selective laser sintering of alumina/PA microspheres. *Ceram Int* 2012;38:1241–7.

[18] Wohlert M, Bourell D. Rapid prototyping of Mg/SiC composites by a combined SLS and pressureless infiltration process. *1996 International Solid Freeform Fabrication Symposium*, 1996.

[19] Mott M, Evans JRG. Solid freeforming of silicon carbide by inkjet printing using a polymeric precursor. *J Am Ceram Soc* 2004;84:307–13.

[20] Rodrigues SJ, Chartoff RP, Klosterman DA, Agarwala M, Hecht N. Solid free form fabrication of functional silicon nitride ceramics by laminated object manufacturing. In: Bourell DL, editor. *2000 International Solid Freeform Fabrication Symposium*, University of Texas at Austin; 2000, pp. 1–8.

[21] Myers K, Cortes P, Conner B, Wagner T, Hetzel B, Peters KM. Structure property relationship of metal matrix syntactic foams manufactured by a binder jet printing process. *Addit Manuf* 2015;5:54–9.

[22] Simon JL, Michna S, Lewis JA, Rekow ED, Thompson VP, Smay JE, et al. In vivo bone response to 3D periodic hydroxyapatite scaffolds assembled by direct ink writing. *J Biomed Mater Res A* 2007;83:747–58.

[23] Deckers J, Vleugels J, Kruth J-P. Additive manufacturing of ceramics: a review. *J Ceram Sci Technol* 2014;5:245–60.

[24] Oh Y, Bharambe VT, Adams JJ, Negro D, MacDonald E. Design of a 3D printed gradient index lens using high permittivity ceramic. *2020 IEEE International Symposium on Antennas and Propagation and North American Radio Science Meeting*, 2020, pp. 1431–2.

[25] Sebastian MT, Ubic R, Jantunen H. Low-loss dielectric ceramic materials and their properties. *Int Mater Rev* 2015;60:392–412.

[26] Petosa A, Ittipiboon A. Dielectric resonator antennas: a historical review and the current state of the art. *IEEE Antennas Propag Mag* 2010;52:91–116.

[27] Reaney IM, Iddles D. Microwave dielectric ceramics for resonators and filters in mobile phone networks. *J Am Ceram Soc* 2006;89:2063–72.

[28] Oh Y, Bharambe V, Mummareddy B, Martin J, McKnight J, Abraham MA, et al. Microwave dielectric properties of zirconia fabricated using NanoParticle JettingTM. *Addit Manuf* 2019;27:586–94. https://doi.org/10.1016/j.addma.2019.04.005.

[29] Sebastian MT. Aliminat Totannia, Ceria, Silica, Tungstate and other materials. *Dielectr Mater Wireless Commun* 2008:379–443. https://doi.org/10.1016/b978-0-08-045330-9.00011-x.

[30] Dreike PL, Fleetwood DM, King DB, Sprauer DC, Zipperian TE. An overview of high-temperature electronic device technologies and potential applications. *IEEE Trans. Compon Packag Manuf Technol Part A* 1994;17:594–609.

[31] Chasserio N, Guillemet-Fritsch S, Lebey T, Dagdag S. Ceramic substrates for high-temperature electronic integration. *J Electron Mater* 2009;38:164–74. https://doi.org/10.1007/s11664-008-0571-8.

[32] Wachtman JB. Mechanical and Thermal Properties of Ceramics: Proceedings. U.S. Department of Commerce, National Bureau of Standards; 1969.

[33] Mummareddy B, Maravola M, MacDonald E, Walker J, Hetzel B, Conner B, et al. The fracture properties of metal-ceramic composites manufactured via stereolithography. *Int J Appl Ceram Technol* 2020;17:413–23. https://doi.org/10.1111/ijac.13432.

9 Considerations for Design and Manufacturing of Flex Devices and Printed Conductive Elements

Ken Blecker
United States Army Combat Capabilities
Development Command – Armaments Center

9.1 INTRODUCTION

Flexible (abbreviated 'flex') and flex-hybrid devices require special consideration in designing for high reliability and power optimization. This chapter includes factors that impact reliability and quality, the impact of material properties on manufacturability, the influence of environment and usage on performance, an overview of several potential lifecycles common to flex and flex-hybrid devices, and a review of material through case studies.

As flexible hybrid electronic devices become more prolific, the importance of power optimization and reduction of losses becomes increasingly important. The power source is a leading driver for replacement of devices and overall device lifecycle. While sensor saturation or calibration also drives a portion of the replacements, these components can be augmented with a backup version more easily and at a lower cost than a second power supply.

Designing for flex electronics is fundamentally different from designing for conventional rigid electronics [1]. Conventional electronics predominantly live on a Euclidean plane or a series of intersecting planes. This presents a limitation in the available form factors and the practical electronic-to-surface interfaces. Flex electronics provide more options for the electronic-to-surface interface and allow for multiple surface geometries to be supported by a single flexible device as shown in Figure 9.1. As the device can conform to multiple surface features and geometries, the applications for a single design are broader [2].

While it is beneficial to have an electronic package that can be deployed to multiple surfaces, it also comes at a price. The performance and inherent noise of the flex electronic or additively manufactured portion of the electronic device is driven by the strain state of the device which will be covered in more detail later. Additionally,

DOI: 10.1201/9781003138945-9

143

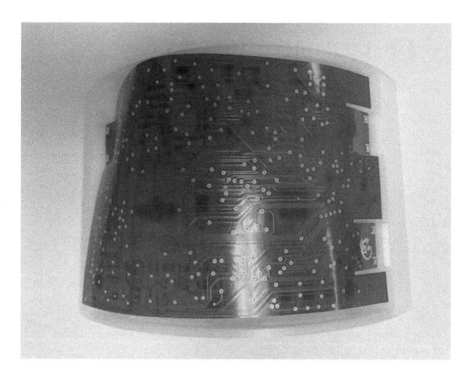

FIGURE 9.1 The flexible substrate can conform to a variety of circumferences.

the mounting methods available to flex or hybrid electronics are different from those available to conventional rigid devices.

The implementation of flex electronics allows for a broader application space and innovative technology with some additional considerations. This includes the ability to link multiple rigid components, conforming to a surface, the potential for weight savings, the potential for material savings, and ease of application.

Reliability and availability are sometimes used interchangeably. In this context, battery life is the main driver for the availability of a device. In order to improve the battery life or reduce the total power requirement for a device, the data collection interval, communication protocol and scheme, and data storage should all be characterized [3]. Flex devices have a more dynamic power usage profile than rigid electronics that is driven by environmental factors like temperature and humidity as well as physical factors like strain or bending. The resistance of flex conductors increases with a decrease in temperature similar to conventional conductors. The resistance of printed conductive traces increases when a flexible substrate is bent or strained [4]. This is because the printed conductive material is not homogenous and instead is composed of many particles or platelets that can shift and separate [5]. Humidity is a secondary issue that can cause some flexible substrates to swell and induce a strain on the components.

While there is a great deal of variability in the performance or internal resistance of flex devices, the fundamental understanding of these phenomenon is not well understood or characterized. Where a designer could refer to a material specification or performance curve for conventional rigid electronics, these curves and characterization activities do not exist on the flex side. Because of the inherent variability in flex design and the novel nature of materials that are being developed and sold and not yet fully characterized, the characterization work is both in its infancy and significantly wider in scope than experienced with rigid materials, components, power sources, and technologies. More specifically, flex electronics do not yet have significant repository of performance and usage data to simulate and estimate power consumption in a variety of usage scenarios where conventional electronics do.

Flex devices are much more sensitive to environmental exposure and cycling than rigid electronics are; factors such as temperature, vibration, and chemical exposure can impact power usage and degradation of flex devices, and because of the more delicate nature of flex devices and often the need to retain that flexible characteristic during their lifecycle, conventional methods of hardening electronics including potting, conformal coatings, and radiofrequency shielding often provide less protection [6] and have pitfalls in the manufacturing and deployment of the technologies that need to be accounted for. An example of a hardened section on a flexible substrate is shown in Figure 9.2.

The manufacturing of flex and flex-hybrid devices is continuing to mature. While it is possible to produce a wide variety of devices and components, it is not possible to produce these devices with as low a trace width variance as is possible with rigid devices [7]. As the characterization at the material level and fundamental understanding of the influence of manufacturing variance on device performance matures [8],

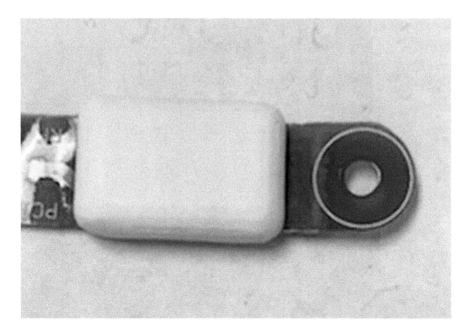

FIGURE 9.2   A ceramic encased component on a flexible substrate providing protection to a discrete portion of the device but also allowing the remainder of the device to retain conformal capabilities.

this manufacturing understanding will evolve and both the procedure for validating manufacturing and the breadth of variability will be characterized.

As the industry standard for manufacturing capability increases for flex devices, the material and substrate sources of supply are also increasing. Because each material and substrate is produced differently and because the basic understanding of the influence of manufacturing variability has not extended to lifecycle or assembly understanding [9], individual producers, lines, or product groups are pursuing an optimal design in different directions. As these optimization routines converge, both a better understanding of the optimal solution as well as how to control variance throughout the lifecycle as a function of material or substrate will continue to strengthen. Multiple methods may be employed on a single device such as that shown in Figure 9.3.

Both battery and interconnects deserve special consideration in flex and flex-hybrid devices. As the understanding of the limitations and prospects of substrates solidifies, the rational for interconnects between different types of substrates will strengthen. As the understanding of the interface performance between conventional, screened, and printed features grows, the rational for selecting and employing a specific interconnect method or device will also mature. Because the interconnect here can be between two or more substrates with different thermal expansion coefficients and

FIGURE 9.3 The connector is through hole mounted, while the remainder of the components are surface mounted. The through hole mounting provides a stronger connection, and the metal reinforcement spreads the strain of plugging and unplugging over a wider area.

different levels of flexibility, the understanding of the performance over the complete lifecycle as well as the robustness of the design and viability of that design will drive reliability at these crossroads.

The choice of primary or secondary cell batteries is also an important consideration. As low-cost and light-weight are driving forces in many flex and flex-hybrid designs, having a light-weight, perhaps rechargeable, power solution is desirable. Having a rechargeable connection can be a concern for a fully encapsulated device. If a plug is required for the recharging connection, then that portion of the device will not be encapsulated and will have a less effective cap or other sealing method. If recharging is accomplished through inductance or other wireless means, then an antenna and associated circuitry will need to be added which can increase size and cost [10]. Having a fully encapsulated device with a primary cell battery does shorten the potential life, but when protection is required, the trade-off of replacing the device at a regular interval to accessing the device with a charger and possibly violating the integrity of the seal must be considered.

Having a flex device that can be applied once, and remain applied throughout its useful life, is desirable. However, considerations such as calibration and battery recharging may limit this approach. For example, some calibrations require the device to be removed and brought back to a laboratory. The removal of the device and replacement of its components have a potential for inducing strain in the device [11] and should be considered as contributing factors for the device's useful lifecycle. Battery recharging has similar issues where the device could either be brought to a charging station or charged in place. If the device is charged in place, then sufficient access to couple the flex device to the charger should be maintained. Figure 9.4 shows a device intended for a single application over the lip of a container.

Internal resistance is a driving characteristic in the development of characterized materials and dialing-in of manufacturing processes. The variability of internal resistance of an electrical interconnect influences both power usage as well as signal-to-noise ratio of sensors of a device. Internal resistance is influenced by usage (stretch, temperature, corrosion), environment, and manufacturing process (sintering, quality of materials, substrate, potting, and interconnects) (Figure 9.5). Reducing the resistance variation over the lifecycle can both reduce the power requirement for the device as well as maintain the required performance such as antenna range, sensor response signal-to-noise ratio, and battery life. In any case, lifecycle power expectations should be defined early and be based on usage scenarios, replacement schedules, calibration time investment, and a detailed concept of operations. As the deployment of internet of things (IoT) and flex in general propagates, the results of power optimization comprise a larger and larger amount of funding and resources.

When a printed conductor stretches, the resistance increases which means the overall power usage of the device increases and more power must be driven through the circuit to provide the same power to the end component or to detect the response of a sensor or discrete component [12]. As the stretching increases, the resistance reaches a maximum, and the resistance may spike and then goes open circuit as the yield limit of the conductive element is reached. While physical controls can be put in place to limit the extent of stretching in many scenarios, this increases the material and cost burden of the device and adds to the weight. Understanding the limits of

FIGURE 9.4   The pictured device is applied over the lip of a container and is not intended to be removed until the end of its life.

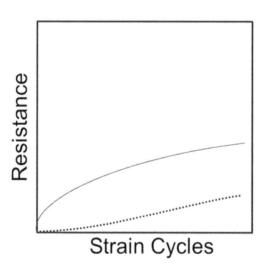

FIGURE 9.5   Resistance as a function of strain cycles, the solid line represents the maxima and the dotted line represents minima of each cycle.

the substrate and traces is an important step in characterizing the performance and durability of the device, as well as being able to relate the power requirements and performance to the usage scenarios identified.

The resistance of conductive traces and flexible components changes dynamically with bending. There is a practical maximum bend angle and maximum number of bends before the trace cracks which must be determined empirically, but there is also a limit relating to the maximum internal resistance of the components that needs to be characterized and considered [13]. Controls can be put in place to limit the extent of bending to prevent cracking, but developing an understanding of the expected design and concept of operation must inform this process. Controls are more difficult to implement for cyclical or high cycle fatigue style bending as well as having a more difficult task in characterizing the material response to such a stimuli. Figure 9.6 shows resistance as a function of strain for printed conductive features.

Folding must be considered both in assembly as well as in use. Folding occurs when the device is assembled in a flat orientation and then changed to a non-planar orientation for application or when its plane status changes during operation as it flexes [14]. Because the device is printed in one plane state and then transitioned to another non-Euclidean state from when it was printed or screened, there are residual stresses. For conductive traces, this is a problem for both work hardening and cracking, which are present in different failure modes. In general, these folded regions will exhibit higher local resistance and may act like an interface or interconnect between two regions of the same material where the crease, crack, or node forms. If assembly folding is the main concern, some moderation of the problem may be possible by folding and then a secondary printing or sintering operation to relieve some of this stress or repairing damage.

9.2 THERMAL CONSIDERATIONS

Thermal expansion of flex and flex-hybrid devices causes disparate expansion of layers or components that can lead to delamination or cracking. Several general thermal exposure scenarios exist such as being mounted to people, mounted to equipment,

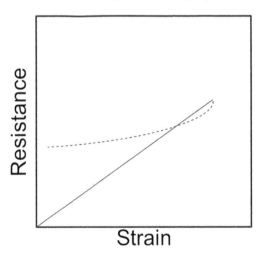

FIGURE 9.6   Printed conductive material is a conglomerate of material suspended in a binder. As this material is strained, the internal connections are changed.

mounted to assets, and the time period prior to application of the device. Figure 9.7 shows a device mounted in a plastic container with cracked potting material.

When mounted to people, the temperature of the mounted side is driven by the temperature of the person or clothing and the opposing side is driven by the environment which can vary from extreme desert to extreme arctic with humidity varying from condensing to near zero. Because most people tend to have the same temperature, accounting for the person side of the device is straight forward. This becomes more complicated when the device is mounted to clothing or other parts of the equipment kit. Because many parts of the kit have varying levels of separation from the outer surface of the skin, varying thermal capacitances, and varying levels of heat generation as a result of power sources or heat rejection due to electronic or other devices, this must be carefully considered and documented prior to design or deployment.

The outward facing side of devices can see a wide variation in temperatures such as ambient conditions and weather varying from −20°F to +135°F (−28.9°C to +57.2°C) with a steep thermal gradient, an office environment varying from 60°F to 80°F, an industrial environment varying from 40°F to 120°F, a field environment varying from −20°F to 135°F, an athletic environment varying from 90°F to 120°F with high humidity, and a medical environment varying from 60°F to 100°F. Each of these temperature gradients occurs across the relatively short thickness of the device and can cause thermal expansion problems both in a steady-state as well as a dynamic condition [15].

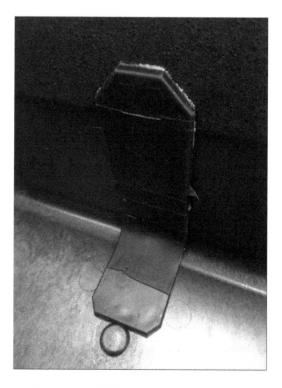

FIGURE 9.7　The potting material of this device mounted on the interior of this container is cracked due to a mismatch of the movement of the container wall and the device.

When mounted to equipment, the environment is often more controlled and consistent. Industrial settings have long periods of constant temperature due to steady-state manufacturing or control operations. Mounting to equipment for transportation assets presents more variability, but periods of exposure at ambient conditions followed by periods of operation at more extreme temperatures should be expected. The temperatures can range from −40°F to 300°F in normal operations. Because of the near constant temperature exposure and well defined usage cases, this is a good candidate for energy harvesting. Figure 9.8 shows an option for overcoming wide thermal exposure ranges.

When a device is mounted to an asset, it may be for tracking, monitoring some aspect of the assets, or monitoring the environment of the asset during its lifecycle. The temperature variation can range from roughly −40°F to 160°F (−40°C to 7.1°C) with daily cycling, though outliers can be outside of this range. This includes storage of assets in an unbiased state; this can include storage in uncontrolled temperatures throughout the world and in direct sunlight [16]. Due to the uncontrolled environment, this also includes diurnal cycling and variations associated with weather and season. It is important to understand every aspect of the lifecycle to appropriately classify the usage environment as well as how the device will be employed. Because of the unknown nature of a particular storage life, it is difficult to plan for energy harvesting or other novel energy scheme.

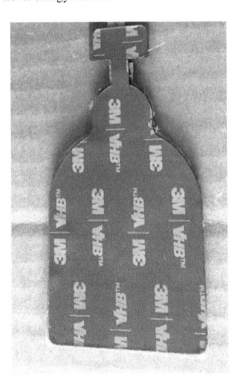

FIGURE 9.8   Foam cored tape allows for mounting a device on a substrate with a thermal expansion coefficient mismatch.

Devices experience thermal exposure prior to being applied to an asset or surface. During this period, the device is usually unbiased or partially biased, and this can cause brown-out or degradation because of lower voltages and high currents. Before the device is applied to a substrate, the power source can degrade and the device itself can exhibit changes in performance because of aging and corrosion. Understanding the concept of stockpiling, managing, and applying devices is important to developing an actionable power optimization plan as well as a viable and cost effective production and storage plan. While there are benefits associated with mass production, these should be balanced with the degradation of safety or performance due to aging.

9.3   MANUFACTURING PROCESS

The manufacturing process and associated quality controls can impact materials and material properties across the device. This includes contact resistance, material loss, internal resistance, material-to-substrate adhesion, and interconnect loss/rigidity.

Contact resistance is a particular problem with flex and printed conductive elements. Because of the high surface area and the presence of binders, the interface between two materials or two passes of the same material can vary greatly within a combination of narrowly controlled parameters such as viscosity, line width, sintering method, and ink pot time [17]. Because of the inherent variability in the process, the internal resistance of a conductive trace or component can also vary. While in situ monitoring of the performance of a component is desirable, this is not always possible or recommended because the transition from ink to sintered or cured trace is also variable. In general, it is important to maximize the surface area of contact between layers or components. This includes selecting materials that will maintain good conductivity with variable contact pressure due to flex or thermal expansion as well as minimize corrosion or galvanic action. It is important to design interconnect to retain contact over the expected range of usage in each scenario; you will consider including testing or evaluating the completed device in-situ where possible.

Material loss is a particular concern as the conductivity is driven by effective contact cross-sectional area and not bulk material properties. Material can be lost because of abrasion between sheets, die cutting, or other operations. Volatile organic compounds (VOCs) or binders can be sintered or slowly leach over time and degrade the surrounding material. When material is lost, both the dust and the detritus are a concern as well as grain structure changes. Figure 9.9 shows a broken connection due to mishandling.

Internal resistance is an item of particular concern. As the conductivity of traces and printed components is driven by the incremental internal resistance of a given cross section, a firm understanding of the processes and resulting product is necessary to control or improve. This includes manufacturing and material documentation and controls. This will begin as capture of information and relation building to the state of the finished product and then move to control of the process with optimization in mind. The information that is important is material characteristics, conductive material to VOC or binder ratio, testing and characterization of completed components, spot testing after sintering or UV cure, and documentation of failures and performance changes.

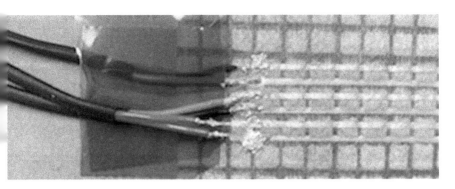

FIGURE 9.9   The connection between the printed components and the conventional wire was broken due to mishandling.

Material to substrate adhesion plays a role in both thermal expansion as well as manufacturing [18]. Maintaining contact between all the materials and the substrate supports durability and continuity. While conductive materials are most often focused on, all materials can lead to a failure or degradation of capability if separation occurs. Additionally, separation creates a void for corrosion or detritus to enter the assembly. Care should be taken to inspect for peeling away, popping off of components, and performance at a component level to accurately identify the failure location.

Interconnects must maintain contact and pressure in a flexible environment, and this often means a rigid island in a sea of flexibility creating a local stress riser. Care should be taken to account for the locally stiff interconnect within the otherwise flexible device. For poorly attached components, this can mean an intermittent open circuit making it difficult to diagnose. Figure 9.10 shows rigid connectors integrated into a flex assembly.

During the design process, it is possible to mitigate some of the risk associated with flex electronics [19]. This can include establishing a max power cut-off for a subsystem or component to account for the increase in internal resistance. Some specific components may see a surge in power usage with a minimal signal output. Degradation in response signal can be driven by the increase in noise due to the increased resistance. It is also possible to include physical constraints to limit the folding, bending, or stretching of the assembled and applied device, though this may also limit the utility and the benefits of going to a flex construct.

9.4   ENVIRONMENT AND USAGE

Here, the focus is on variability to the strain state of the devices by outside influences. When a flexible device is under strain, the attached conductive element can experience an increase in the internal resistance. This is illustrated in Figure 9.11. As the strain increases, the internal resistance increases. This results in both increased power draw for components as well as dynamic signal changes under otherwise stable conditions.

FIGURE 9.10 The connectors are attached to the flexible substrate and slots are cut to allow for multiple directions of flexibility without damaging the surrounding components.

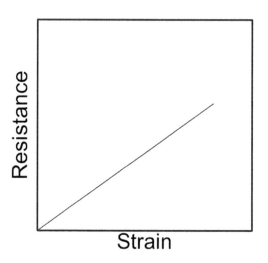

FIGURE 9.11 Theoretical resistance as a function of strain.

As strain changes to stretch and the concern shifts to not just a static state but a dynamic state with recovery, this becomes more complicated. As the strain is reduced, the resistance could return to the baseline or it could recover to come to higher resistance state as dictated by material properties and strain level. Because of the complex and dynamic nature of this change, efforts should be taken to characterize this in an operational environment.

Over time, this stretch and recovery cycle can span a range of potential resistance levels. While it is convenient to assume that the device operates only at the extremes of the resistance plot, it actually operates in the region bounded by the high and low resistance lines providing a significant range or potential states to account for. Over time, the resistance generally increases, and a point will be reached where the conductive elements crack or otherwise break down as this fatigue limit is achieved.

Over time, temperature exposure can also induce a cyclic strain state in addition to potential changes in resistance just due to the temperature of the material. Efforts should be taken to minimize the range of temperature (less than 20°F) that a device will see in operation including mounting to a consistent temperature location or adding an insulating layer to isolate from ambient conditions. At lower temperatures (below 30°F), this can be complicated by cracking, delamination, separation, or shrinkage, and at higher temperatures (above 180°F), this can include melting of adhesives, unmatched thermal expansion, or increased material interactions such as corrosion, galvanic action, and general chemical reactions keeping in mind that printed conductive traces have a very high printed metal surface area and that at the nano-scale many reactions occur that would otherwise require activation energy to kick-off.

Corrosion is a concern not just in thermal cycling but throughout the lifecycle of a device. Flex devices present a particular problem in that cracks and separation occur at a higher frequency that are found in rigid devices. These voids allow for detritus, moisture, and corrosion-inducing materials to enter the assembly and degrade materials, adhesive joints, and interconnects. An unprotected component can locally corrode, such as in a case where it is unpotted or unencapsulated, similar to a humidity sensor where the active area needs to be exposed to ambient conditions to collect data, and serve as an ingress point for further contaminants to enter. In general, corrosion of galvanic degradation in performance eventually results in a full failure of the device. Figure 9.12 shows an encapsulated device with an open section for a sensor in the middle.

Abrasion can occur both during manufacturing and during the lifecycle, and while the physical result is similar, the ability to detect and account for degradation due to abrasion is different. During manufacturing, abrasion can be inspected for and additional material or encapsulation applied. During the lifecycle, abrasion can result in a poor performing device or outright failure. Resistance is a function of a cross-sectional area of a conductor, and as the conductor is worn away, the relative resistance will increase as the conductive cross-sectional area decreases. During the lifecycle, this is often a very small removal of material at a time, but as these cycles aggregate, this removal is significant. If monitoring of the device or sensor performance is possible, this can be detected as a continual change over time. An end state of total failure is almost always the result of this type of degradation.

FIGURE 9.12　The device is encapsulated but a section is left clear to allow for the operation of a sensor in the center of the element pictured.

9.5　LIFECYCLE CONSIDERATION

When designing for the lifecycle, it is important to consider the frequency of replacement or reapplication, the time and effort for calibration, the frequency of interrogation or reporting, and the frequency of data download. The frequency of replacement or reapplication can drive the total onboard power requirement as well as the durability required for the device [20]. The time and effort for calibration impacts the quality of components chosen as well as the reapplication scheme. The frequency of interrogation impacts the total power needed as well as the communication method chosen. The frequency of data download impacts the total memory required onboard, the power required on board, and the interface for data download or the potential for external power to be applied and the onboard power restored.

Reducing complexity of the design of the physical layout of devices has several benefits. First and foremost is reducing the power consumption of the device. While it is not often possible to increase the total power onboard because of volume or weight constraints, reducing the power draw can have significant benefits. Additionally, minimizing the complexity can reduce the opportunities for device degradation or other types of failure. As the complexity of a device increases more considerations must be incorporated into the design to control for potential failure modes. By having a less complex device the application itself can drive the majority of the design. Last, simplifying the device can be a catalyst for improving the design. A careful look at the capabilities and functions can lead to identification of non-value added components and capabilities that can be pruned. In general, it is important to isolate the rigid portion or rigid components from the flexible portion, protect the entire device from the environment, minimize battery changes as the labor cost can outweigh any savings, and evaluate energy harvesting.

9.6   CASE STUDIES

Four case studies will be considered to illustrate the potential design consideration and pitfalls of a flex roll out. These case studies will be a human applied biomonitor, an industrial process controller, a unique identification device for an item in storage, and a flex device embedded in a composite. Each of these case studies has a particular set of concerns and application intricacies.

Designing for a human applied biomonitor is a complex problem. The device has to account for tens of thousands of flex events per day and must remain adhered to skin or fabric over the intended usage period. Because of this high rate of flex, and the potential to capture each flex event, the power requirement for the life of the device should be minimized and optimized. Determination should be made if the device will be replaced on a regular interval by the wearer or if a technician will need to perform the application. Adhesive physical securement, while convenient and flexible in application, may not be able to account for the required power consumption or resistance changes over time. Several special characteristics are present such as sweat, dead skin, and detritus, which can each interact with the device and the adhesive. Additionally, a wireless communication scheme employed at 433 MHz can experience significant interference from water, which is a primary human component.

There is some opportunity for energy harvesting, but the usage case should be carefully drawn out to account for the ambient temperatures that the outward facing side of the device will experience as well as the user experience. Primary batteries are a potential solution, but this could result in a great investment in batteries and assembled device if the batteries are not replaceable as weight and volume are both paramount to the user experience. There is an opportunity to hardwire the sensor to a power source if the usage is in a clinical environment. The communication protocol is generally short range as a phone or device is commonly at hand, while sensor data only can be passed and the device may be keyed to a particular patient or human. Mesh network support is desirable but also comes at a power requirement cost. The motivation for improvement in the human applied monitor space is reducing the tangle of monitoring wires, providing better data, and enabling health care providers.

Designing for an industrial process controller, while the environment is generally well characterized and understood, also presents unique challenges and risks. It is important to define the environment in which the device will be operating in. Many aspects of the device and manufacturing process can be tailored to specific temperature and vibration environments. While these are a design consideration, they are also an opportunity for energy harvesting over the life of the device. Energy storage is often secondary for industrial monitoring devices as a hardwired power source can be umbilicalled from the underlying machine. Where hardwired power is not available, energy harvesting should be considered to save changing or charging batteries and the respective labor activities and costs. Communication protocols should support 10–100 feet of standoff as well as two-way sensor data, machine identification, and control commands. Both inductive and capacitive machine noises are common in this environment and should be studied prior to design or implementation. Supporting a mesh network protocol is often desired and the ability to monitor a location or node that was not previously accessible by sensors is a motivating factor.

Items in storage face a bleak and lonely existence. These items are often placed into storage shortly after production and can remain there for much longer than they would last in use. There is also the need to account for unknown ambient exposure as the storage area may not be environmentally controlled or protected. Understanding the procedure for storage and selection of items is important as a first in, first out method may not be used. Also, determine if the device will provide value during storage, during the operational life of the product, or both. Because of the unknown ambient environment of the device, it is difficult to design an energy harvesting device for this case, and primary batteries are often utilized as they offer a longer absolute lifetime than secondary cells. Communication protocols should span the area of the expected warehouse and should support sensor and asset data as well as batch interrogation. The motivation for implementation of an asset monitoring device is automated inventory and item state, readiness of assets, and to enable automated pick and place of material handling equipment.

Items embedded in composites are becoming more common in both industrial process controls as well as for robust deployment of electronic devices. Because the devices are embedded in composites, it is very difficult to replace or recharge batteries or repair the device. During the design phase, care should be taken to ensure that the device and assembled composite are free of defects and have been tested to account for all known states and failures. Fully encasing or encapsulating a device makes energy harvesting difficult as the thermal gradient within the encapsulation will be small and the encapsulation often provides insulation from vibration effects. Communication protocols are generally limited to the near field and wireless charging is a possibility. The motivation for this case is to support sensor and IoT in a hardened environment, to monitor the integrity of a composite, and to provide a "black box" capability for the underlying asset.

The power usage and performance of a device is a function of manufacturing, application, and usage scenario. Power can be optimized but should be considered in conjunction with cost. While it is difficult to optimize power usage or device design across multiple scenarios, several designs may be required, each individually optimized for a particular scenario. Characterizing materials and interfaces for a particular device and scenario is crucial to understand the power and performance implication of a design choice or implementation concept of operation.

9.7 CONCLUSION

Several important considerations in designing for high reliability and power optimization in flex-hybrid electronic assemblies were identified. While the focus of this chapter was on reliability, quality, manufacturability, and performance, other factors should be considered as well based on the specific usage scenario identified for a particular device. Key to implementation is acknowledging that designing for flex and hybrid electronic architectures is fundamentally different from conventional electronics and should be considered a new design and not an iteration of an existing conventional design. A change in performance or lifetime is also expected as a design is transitioned to flex or hybrid, and the manifestation of this degradation should be identified and characterized to align accelerated testing. Power sources and interconnects present a unique application specific challenge for flex devices, and careful design is key to success.

ACKNOWLEDGMENTS

This chapter was created with the support of Dr. Pradeep Lall of the CAVE3 NSF UICRC and Dr. Giuseppe Di Benedetto of DEVCOM-AC.

REFERENCES

1. Whitt, R., Huitink, D., Emon, A., Deshpanade, A., Luo, F., "Thermal and electrical performance in high-voltage power modules with nonmetallic additively manufactured impingement coolers," *IEEE Transactions on Power Electronics*, vol 36, number 3, 2021, pp. 3192–3199.
2. Schwartz, D.E., Rivnay, J., Whiting, G.L., Mei, P., Zhang, Y., Krusor, B., Kor, S., Daniel, G., Ready, S.E., Veres, J., Street, R.A., "Flexible hybrid electronic circuits and systems," *IEEE Journal on Emerging and Selected Topics in Circuits and Systems*, vol 7, number 1, 2017, pp. 27–37.
3. Cordella, M., Alfieri, F., Clemm, C., Berwald, A., "Durability of smartphones: a technical analysis of reliability and reparability aspects," *Journal of Cleaner Production*, vol 286, 2021, pp. 125–388.
4. Geyik, C.S., Mekonnen, Y.S., Zhang, Z., Aygun, K., "Impact of use conditions on dielectric and conductor material models for high-speed package interconnects," *IEEE Transactions on Components, Packaging and Manufacturing Technology*, vol 9, number 10, 2019, pp. 1942–1951.
5. Malik, M.H., Grosso, G., Zangl, H., Binder, A., Roshanghias, A., "Flip chip integration of ultra-thinned dies in low-cost flexible printed electronics; the effects of die thickness, encapsulation, and conductive adhesives," *Microelectronics Reliability*, vol 123, 2021, pp. 114–204.
6. Gjokaj, V., Papapolymerou, J., Albrecht, J.D., Chahal, P., "Design and fabrication of additively manufactured hybrid rigid-flex RF components," *IEEE Transactions on Components, Packaging and Manufacturing Technology*, vol 9, number 4, 2019, pp. 779–785.
7. Tong, G., Zhou, J., Chang, J., "Flexible electronics: review and challenges," *IEEE International Symposium on Circuits and Systems (ISCAS), 18200521*, 2018.
8. Lall, P., Soni, V., Miller, S., "Effect of U-flex-to-install and dynamic U-flexing on Li-ion battery SOH degradation subjected to varying fold orientations, folding speeds, depths of charge, C-rates and temperatures," *Journal of Electronic Packaging*, Published Online, https://doi.org/10.1115/1.4052750, 2021.
9. Lall, P., Narangaparambil, J., Leever, B., Miller, S., "Flexure and twist test reliability assurance of flexible electronics," *Journal of Electronic Packaging*, vol 142, number 3, 2020, pp. 031121.
10. Syrovy, T., Kazda, T., Akrman, J., Syrova, L., "Towards roll-to-roll printed batteries based on organic electrodes for printed electronics applications," *Journal of Energy Storage*, vol 40, 2021, pp. 102–680.

11. Behfar, M., Khorramdel, B., Korhonen, A., Jansson, E., Leinonen, A., "Failure mechanisms in flip-chip bonding on stretchable printed electronics," *Advanced Engineering Materials*, published online, https://doi.org/10.1002/adem.202100264, 2021.
12. Stewart, B., Ginga, N., Sitaraman, S., "Biaxial inflation stretch test for printed electronics", *2020 IEEE 70th Electronic Components and Technology Conference (ECTC)*, 2018. https://doi.org/10.1109/ECTC32862.2020.00178.
13. Aga, R.S., Kreit, E.B., Dooley, S.R., Devlin, C.L., Bartsch, C.M., "In situ study of current-induced thermal expansion in printed conductors using stylus profilometry," *Flexible and Printed Electronics*, vol 1, 2016, pp. 012001.
14. Lall, P., Soni, V., Miller, S., "Flex-to-install application performance of power sources subjected to varying fold orientations, C-rates, and depths of charge," *2020 19th IEEE Intersociety Conference on Thermal and Thermomechanical Phenomena in Electronic Systems (ITherm)*, 2020. https://doi.org/0017810.1109/ECTC32862.2020.00178.
15. Makita, T., Nakamura, R., Sasaki, M., Kumagai, S., Okamoto, T., Watanabe, S., Takeya, J., "Electroless-plated gold contacts for high performance, low contact resistance organic film transistors," *Advanced Functional Materials*, vol 30, number 39, 2020, pp. 2003977. https://doi.org/10.1002/adfm.202003977.
16. Gajadhur, M., Regulska, M., "Mechanical and light resistance of flexographic conductive ink films intended for printed electronics," *Dyes and Pigments*, vol 178, 2020, pp. 108381.
17. Khan, Y., Thielens, A., Muin, S., Ting, J., Baumbauer, C., Arias, A.C., "A new frontier of printed electronics: flexible hybrid electronics," *Advanced Materials*, vol 32, number 15, 2020, pp. 1905279.
18. Maurya, D., Khaleghian, S., Sriramdas, R., Kumar, P., Kishore, R.A., Kang, M.G., Kumar, V., Song, H., Lee, S., Yan, Y., Park, J., Taheri, S., Priya, S., "3D printed graphene-based self powered strain sensors for smart tires in autonomous vehicles," *Nature Communications*, vol 11, 2020, pp. 5392.
19. Ouchen, F., Aga, R., Harvey, M., Heckman, E., "Space survivability for printed electronics applications," *Flexible and Printed Electronics*, vol 6, number 1, 2021, pp. 015012.
20. Vasara, A., Hakola, L., Valimaki, M., Vilkman, M., Orelma, H., Immonen, K., Torvinen, K., Hast, J., Smolander, M., "Beyond flexible towards sustainable electronics," *SID Symposium Digest of Technical Papers*, vol 52, number1, 2021, pp. 764–767.

10 Three-Dimensional Functional RF Devices Enabled through Additive Manufacturing

Austin Good, Zachary Larimore, and Paul Parsons
Samtec

The geometrical freedoms enabled by additive manufacturing have made it a viable manufacturing method for fabricating functional radiofrequency (RF) devices and systems. In some instances, it is the most viable manufacturing method for geometrically complex RF devices. Secondly, additive manufacturing enables rapid prototyping and extreme customization of RF devices without significant retooling costs and lead times. In this chapter, we will describe several RF systems that demonstrate (1) the additive manufacturing capability to fabricate unique devices that are incredibly challenging to fabricate through traditional manufacturing processes, and (2) the additive manufacturing's ability to provide rapid fabrication/prototyping with diverse customization of RF characteristics.

10.1 INTRODUCTION: ADDITIVE AS THE MOST VIABLE MANUFACTURING METHOD

One of the more promising applications of AM for RF devices and structures is for the fabrication of substrates, consisting of a single or multiple materials, into a structure with heterogeneous material properties. For example, to design a transmission line element with minimal loss, a designer may desire a substrate with a low dielectric constant, where the transmission line may feed a radiating element such as a patch antenna. However, there may be spatial constraints on the overall system level that restrict the available working area of the antenna, requiring the use of higher permittivity substrate materials for the antenna rather than ideal conditions for the feeding network. Historically, designers of this RF system were left with a choice, either compromise on the performance of the transmission line or compromise on the performance of the antenna. This example demonstrates the limits of conventional manufacturing, and how AM becomes an increasingly attractive and viable manufacturing method. Using traditional manufacturing methods, single substrates with spatially varying material properties are either not possible or very difficult to make, despite not necessarily being geometrically complex. However,

DOI: 10.1201/9781003138945-10

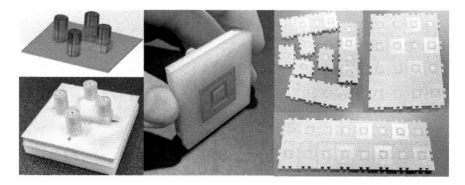

FIGURE 10.1   Dual-band multi-material antenna array printed monolithically through AM.

multi-material additive manufacturing is ideally suited to producing functionally graded substrates such as this. The layer-by-layer and voxel-by-voxel approach of the AM process allows the designer and fabricator the freedom and flexibility to strategically place specified material properties in any given geometric location [1–6]. An example of utilizing this capability in a complex antenna structure is shown in Figure 10.1. This structure is a dual-band, nested, stacked patch antenna array [5,6]. To realize this device, three different dielectric constants must be discretely realized throughout the substrate and superstrate, while maintaining a compact, single structure. Fabricating this, along with the metallic elements which are both in and out of plane, is very difficult, time-consuming, and requires several pieces of equipment if traditionally manufactured. Using AM processes, this entire structure can be printed monolithically on a single system, including the metallic elements. This device was printed on an nScrypt 3Dn-300 using both custom formulated as well as commercial off-the-shelf materials.

Another application where AM is the superior method of manufacturing is in graded dielectric or graded index (GRIN) structures. Similar to the previously described antenna substrate, being able to locally control the permittivity of a material can result in unique performance characteristics, and AM is one of the few manufacturing methods that has been demonstrated as a viable route to producing complex 3D structures with voxel-level control of local permittivity. This is primarily accomplished by tailoring the geometry of a structure, such that the local fill fraction is varied in a controlled manner at any given point in space at a sub-wavelength level. This sub-wavelength mixture of air and polymer results in a local permittivity that can be controlled and predicted using effective medium approximations, resulting in a near-continuously graded structure. One common application of this grading method is graded index lenses, the most common example being the Luneburg lens [7–18]. The Luneburg lens has a permittivity distribution that yields desirable beamforming characteristics, and through beam switching, a beam can be rapidly switched and redirected in a desirable direction. However, fabricating the permittivity distribution is nearly impossible through traditional manufacturing methods. The most common ways that have been explored for producing these structures through traditional means involve either: (1) drilling holes in thin plates in the desired pattern and then stacking those plates to realize the 3D nature required [8,9], or (2) casting core-shell concentric

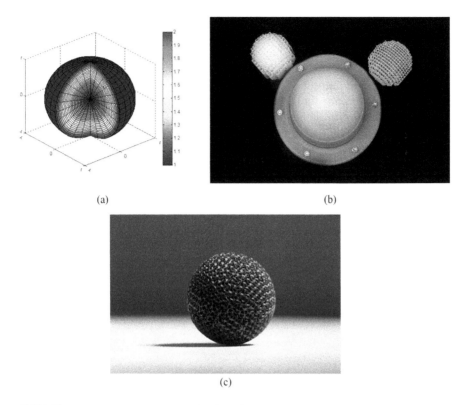

(a)   (b)

(c)

FIGURE 10.2   (a) Permittivity distribution of GRIN lens, Luneburg Lens. (b) Luneburg lens fabricated through FDM [12–16]. (c) Fortify Luneburg lens fabricated through DLP [19].

structures [10]. The drawbacks of the former are that the sheets end up being so significantly perforated that they lose all structural integrity, the sheets still require bonding which can fill the drilled voids altering the desired permittivity distribution, and this method tends to result in undesired anisotropy in the stacking vector. The primary issues relating to the latter traditional manufacturing method are that there are limited materials available with discrete permittivity values which results in more of a step-wise permittivity distribution than desired, and that any air pockets between cast layers distort the permittivity profile, leading to poor performance. Additive manufacturing through Fused Deposition Modeling (FDM), photopolymer systems, and polymer jetting has proven to be the most viable manufacturing methods for realizing Luneburg-like permittivity distributions (Figure 10.2).

10.2 ADDITIVE PROVIDES RAPID PROTOTYPING AND EXTREME CUSTOMIZATION OF RF DEVICES

Additive manufacturing allows the user to transition from design to fabrication of functional devices/prototypes, and from testing through evaluation at rates that traditional manufacturing typically cannot match. This is largely due to AM's ability to fabricate functional structures without significant investment in tooling and with minimal waste. For commercial off-the-shelf (COTS) RF vendors, such as Pasternack, Eravant, and Fairview Microwave, the consumer is locked into pre-defined specifications best suited for a majority of their consumer base. However, if the consumer requires an antenna with a non-standard frequency range or radiation pattern, it is non-trivial to find a design and fabrication house capable of producing said custom antenna through traditional manufacturing needs. However, if the consumer is capable of defining desired RF characteristics, AM is well suited to the production of these types of parts [20–22]. The following real-world case study describes this scenario, from concept, design, fabrication, to validation.

Illustrated in the figures below, a customer needed a custom radiation pattern, as shown in Figure 10.3 with −30 dB of cross-polarization isolation across a majority of the X-band. This needed to be completed within a month. The most turn-key approach was an additively manufactured orthomode transducer to achieve

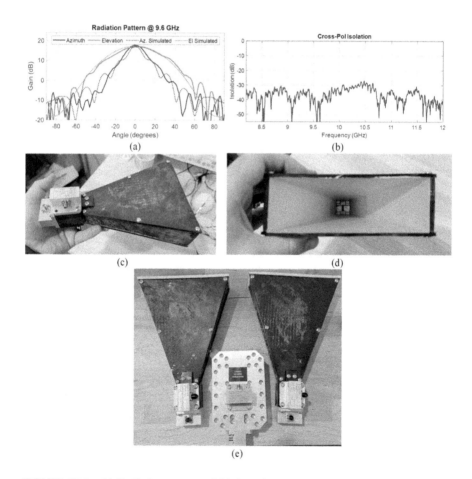

FIGURE 10.3   (a) Radiation pattern of fabricated custom antenna, (b) isolation between antenna ports, (c) side view of antenna, (d) opening view of antenna, and (e) top-down view of antennas next to COTS orthomode transducer.

high cross-polarization isolation with a pyramidal horn antenna to achieve desired radiation pattern. The RF design of the system was completed in a Finite Element Modeling (FEM) software within a week [23]. Fabrication of the antenna and ortho-mode transducer was done through FDM with COTS thermoplastic, and metallization of parts was done with COTS metal paste. Several design cycles were required to optimize performance (which additive manufacturing excels at). The final design is shown in Figure 10.3, matches the desired performance, and was completed in 2 weeks. This expedited timeline while matching custom RF performance specifications showcases additive manufacturing's ability to provide rapid prototyping and extreme customization of RF devices.

REFERENCES

[1] Mitra, Dipankar, et al. "A fully 3D-printed conformal patch antenna using fused filament fabrication method." *2020 IEEE International Symposium on Antennas and Propagation and North American Radio Science Meeting*. 2020.

[2] Sorocki, Jakub, et al. "Realization of compact patch antennas on magneto-dielectric substrate using 3D printing technology with iron-enhanced PLA filament." *2018 International Conference on Electromagnetics in Advanced Applications (ICEAA)*. 2018.

[3] Mitra, Dipankar, et al. "A 3D printed microstrip patch antenna using electrifi filament for in-space manufacturing." *2021 United States National Committee of URSI National Radio Science Meeting (USNC-URSI NRSM)*. 2021.

[4] Ramly, Athirah Mohd, et al. "Design of a circular patch antenna for 3D printing." *2016 International Conference on Computer and Communication Engineering (ICCCE)*. 2016.

[5] Mitchell, Gregory. "A nested antenna approach for a quad-band dual polarization antenna element." *2021 IEEE-APS Topical Conference on Antennas and Propagation in Wireless Communications (APWC)*. 2021.

[6] Mitchell, Gregory, Zachary Larimore, and Paul Parsons. "Additive manufacturing of a dual band, hybrid substrate, and dual polarization antenna." *2020 International Applied Computational Electromagnetics Society Symposium (ACES)*. 2020.

[7] Luneburg, Rudolf K. *"Mathematical Theory of Optics"*. University of California Press, New York, 1966.

[8] Johari, Esha, Zubair Akhter, and M. Jaleel Akhtar. "Design of the modified cylindrical Luneberg lens antenna for millimeter wave imaging." *2015 IEEE International Symposium on Antennas and Propagation & USNC/URSI National Radio Science Meeting*. 2015.

[9] Rondineau, Sébastien, Mohamed Himdi, and Jacques Sorieux. "A sliced spherical Luneburg lens." *IEEE Antennas and Wireless Propagation Letters* 2 (2003): 163–166.

[10] Bor, Jonathan, et al. "Light and cheap flat foam-based Luneburg lens antenna." *The 8th European Conference on Antennas and Propagation (EuCAP 2014)*. 2014.

[11] H. Xin and M. Liang, "3-D-printed microwave and THz devices using polymer jetting techniques." *Proceedings of the IEEE* 105.4 (2017): 737–755. https://doi.org/10.1109/JPROC.2016.2621118.

[12] Larimore, Zachary, et al. "Use of space-filling curves for additive manufacturing of three dimensionally varying graded dielectric structures using fused deposition modeling." *Additive Manufacturing* 15 (2017): 48–56.

[13] Regazzoni, Carlo S. and Andrea Teschioni. "A new approach to vector median filtering based on space filling curves." *IEEE Transactions on Image Processing* 6.7 (1997): 1025–1037.

[14] Z. Larimore, S. Jensen, A. Good, A. Lu, J. Suarez and M. Mirotznik, "Additive manufacturing of Luneburg lens antennas using space-filling curves and fused filament fabrication." *IEEE Transactions on Antennas and Propagation* 66.6 (2018): 2818–2827.

[15] Biswas, Soumitra, et al. "Realization of modified Luneburg lens antenna using quasi-conformal transformation optics and additive manufacturing." *Microwave and Optical Technology Letters* 61.4 (2019): 1022–1029.

[16] Liang, Min, et al. "An X-band Luneburg lens antenna fabricated by rapid prototyping technology." *2011 IEEE MTT-S International Microwave Symposium.* 2011.

[17] Liang, Min, et al. "A 3-D Luneburg lens antenna fabricated by polymer jetting rapid prototyping." *IEEE Transactions on Antennas and Propagation* 62.4 (2014): 1799–1807.

[18] Sayanskiy, Andrey, et al. "Broadband 3-D Luneburg lenses based on metamaterials of radially diverging dielectric rods." *IEEE Antennas and Wireless Propagation Letters* 16 (2017): 1520–1523.

[19] "Low Loss Dielectric Printing. https://3dfortify.com/wp-content/uploads/2022/01/RF-One-Pager_RevF.pdf

[20] Toy, Yunus Can, et al. "Design and manufactering of an X-band horn antenna using 3-D printing technology." *2017 8th International Conference on Recent Advances in Space Technologies (RAST).* 2017.

[21] Wahyudi, Agus Hendra, et al. "3D printed wideband circularly polarized pyramidal horn antenna with binomial polarizer for CP-SAR application." *2020 IEEE Asia-Pacific Microwave Conference (APMC).* 2020.

[22] Vaske, Anne, Altan Akar, and Björn Neubauer. "Coplanar waveguide fed U-band horn antenna manufactured using 3D printing and electroplating." *2022 16th European Conference on Antennas and Propagation (EuCAP).* 2022.

[23] U. Rashid, "Design and simulation of orthomode transducer in Ku-frequency band on HFSS." *2008 IEEE International Multitopic Conference*, Karachi, Pakistan, 2008, pp. 290–293, https://doi.org/10.1109/INMIC.2008.4777751.

Index

Note: **Bold** page numbers refer to tables and *italic* page numbers refer to figures.

abrasion 152, 155
acrylonitrile butadiene styrene (ABS) 8, 9, 51, 106
actuation, electro-mechanical 100–101
actuators 93
  electro-mechanical 100
additive manufacturing (AM) 1, 4, 31, 104, 115
  for antennas 31–34
  characteristics of 47
  defined as 46
  to fabricate ceramic parts **130**
  for fabricating advanced pressure sensors *59, 59–60, 60*
  high-temperature antenna *36*
  for hypersonics 35
  metal-based 69
  polymer-based 71
  power of 38
adhesion 7
adhesion evaluation 12
  of aerosol jet conductive inks 20–21, *21*
  of micro-dispensed conductive inks 13–14, *14*
adhesion tool 12
aerosol jet conductive inks 18, 20
  adhesion evaluation of 20–21, *21*
  electrical interconnect testing with 21–22
  vs. micro-dispensed conductive inks **20**
aerosol jet printing (AJP) 33, 95, 105
alumina 36, 132
Ambit™ tool 59
antennas
  additive manufacturing for 31–34
  all-metal meta surface 38, 39
  for conical and cylindrical geometries 37
  CubeSat 38
  high-temperature 35–37, *36*
  for high-temperature and hypersonic applications 34–37
  for hypersonic applications 35, 37
Antimony Tin Oxide (ATO), on 3D-printed zirconia 134
applications
  hypersonic 34–37
  inductors and wireless power 80–84, *81*
artificial intelligence (AI) 24
ASTM F2792 Standard Terminology for Additive Manufacturing Technologies 46
atmospheric pressure plasmas 116

Atomic Force Microscopy (AFM) indentation technique 99
Au clusters, custom-patterned implantation of 100

barium titanate (BaTiO$_3$), piezoelectric nanoparticles of 103
binder jetting (BJ) process 33
biodegradable cellulose acetate (CA) sheets 96
biomonitor 157
biosensors, electrochemical microfluidic 105

ceramics 46, 129, 131
  3D printing process of 129–130
ChemCubed Ag1037 ink 20, 21
coefficient of thermal expansion (CTE) 133
commercial off the shelf (COTS)
  components 4
  metal paste 165
  RF vendors 164
communication protocols 158
complementary metal oxide semiconductor (CMOS) 48
components
  commercial off the shelf 4
  copper-plated AM microwave 39
  radiofrequency 32
  resistance of conductive traces and flexible 149
composite materials, conductive 71
computational fluid dynamics (CFD) simulations 34
computer-aided design (CAD) model 47
computer numerically controlled (CNC) machining 38
conductive composite materials 71
conductive inks
  copper 18
  DuPont CB028 micro-dispensed 23
  printable 96
  silver 18, 33
conductive materials 50
conductive surface layer 71
conductive traces 149
  resistance of 149
  on transparent flexible substrates 96
conical 12-element phased array 37
conical geometries, antennas for 37

contact resistance 152
conventional electronics 143
coplanar waveguides (CPW) 36
copper
    conductive ink 18
    deposition of bulk metals 120–123, *121–123*
    strain gauge structure of 125, *126*
copper-plated AM microwave components 39
corrosion 155
CubeSat antennas 38
custom code, generation 5
cylindrical geometries, antennas for 37

Defense Advanced Research Projects Agency
    (DARPA) 115
defense systems, next-generation 60
deposition of bulk metals
    copper 120–123, *121–123*
    zinc 123–125, *124, 125*
dielectric elastomer actuators (DEA) 99
dielectric rod antennas (DRA) 36
Digital Light Processing (DLP) unit 132
directed energy deposition (DED) 59–60
direct ink writing (DIW) of silver inks 71
direct-print additive manufacturing (DPAM) 33,
    36, 37
direct write plasma-based techniques 116–120,
    119, 120
direct write printing technologies 115
drop-on-demand (DOD) mode 56
DuPont CB028 micro-dispensed conductive
    ink 23

elastomeric/stretchable materials, conductive
    traces implanted in 96–100
electrical interconnect testing 23
    with aerosol jet conductive inks 21–22
    with micro-dispensed conductive inks **16,**
        16–18, *17,* **19**
electrical resistance 76, 83, 105, 108
electric discharge machining (EDM) technique
    118
electrochemical microfluidic biosensors 105
electrochemical techniques 70
electroless plating 70, 83
    metallization 70, 72–74, *73*
electrolytic double-layer capacitors (EDLCs) 104
electromagnetic test, 3D printed ZrO_2 *132*
electro-mechanical actuation 100–101
electronics printers, commercial tabletop 4–5
electroplating, metallization 71
energy harvesters 93
environmental scanning electron microscope
    (eSEM) 121
environment and usage, flex and flex-hybrid
    devices 153, *154,* 155
Euler's analysis 49

eutectic gallium–indium (eGaIn) 74
evaporation-based deposition, metallization 79

fabricating advanced pressure sensors, AM
    methodologies for *59,* 59–60, *60*
fabrication of sensing devices 48
femtosecond laser ablation (FSLA) 99
fiber Bragg grating (FBG) 51
fiber encapsulation additive manufacturing
    (FEAM) 51
filament, selection for fused deposition modeling
    49–50
Finite Element Modeling (FEM) software 165
flex conductors 144
flex devices 144, 145
    thermal considerations 149–152, *150, 151*
flex electronics
    designing for 143
    implementation of 144
flex-hybrid devices, thermal considerations
    149–152, *150, 151*
flexible electrodes 96
flexible hybrid electronic devices 143, 146
fused deposition modeling (FDM) 48
    filament selection for 49–50
    printed pressure sensors 50–51, *52*
fused deposition of ceramics (FDC) 49
fused filament fabrication (FFF) 33, 34, 71, 95
    AM technique of 105

galinstan 74
gas flow, simulation of *120*
GIX Microplotter 57
graded index (GRIN) 162
    lens *163*
Green state Alumina-ATO LTCC concept *140*

halloysite nanoclays (HNCs) 102
harsh/extreme environment 2, 4, 34, 47, 60, 100
    AM antennas used for 32
    evaluation of adhesion, RF performance, and
        interconnects 13–23
    lack of resiliency in 115
    robust and survivable electronics in 69
high g testing 11–14, *12,* 16, 18, 21, 22
high-temperature antennas 35–37, *36*
high-temperature electronic devices 133
high-temperature environment 35
high-temperature sensors 35–37, *36*
hydraulic systems 133
hydrogen gas 117
hypersonic applications, antennas for 37
    antennas for 34–37
hypersonics, AM for 35

inductors and wireless power 80–84, *81*
Industrial Revolution of the Digital Age 4

ink 71
  adhesion, testing protocols 6–7
  ChemCubed Ag1037 20, 21
  direct ink writing of silver 71
  DuPont CB028 micro-dispensed conductive 23
  platinum ink 36
  printable conductive inks, formulation of 96
  silver-particle-based 36
inkjet-printed pressure sensor 57, *58*
inkjet printing (IJP) 56–57
interconnects 2, 4, 5
interdigital electrode (IDE) 54
internal resistance 147, 152
ionic liquids (ILs) 100
ionogel
  metallization of 100–104, *101, 103*
  photo-polymerization of 102
ionogel-electrode interface 104
ionogel/metal electro-mechanical soft actuators, operational principle of *101*
isopropyl alcohol (IPA) 134

Ka-band antenna 37, *38*

laser beam melting (LBM) 33
laser enhanced direct printing additive manufacturing (LE-DPAM) 33
laser metal deposition (LMD) 56, *56*
laser powder bed fusion (LPBF) 33, 39
lifecycle consideration, flex and flex-hybrid devices 156–157
light-emitting diodes (LEDs) 56, 106
light interferometry of copper 121, *121*
liquid injection-based approaches 76
liquid metal
  inductors 82
  metallization 74–76, *75*
Lithium Lanthanum Zirconium Oxide (LLZO) 124, *125*
low Earth orbit (LEO) missions 37
low-temperature co-fired ceramics (LTCC) 37, 138
Luneburg lens 31, 162

machine learning (ML) 24
manufacturing process, flex and flex-hybrid devices 152–153, *154*
Mars Cube One (MarCO) CubeSat mission 39
MasterBond EP21 epoxy resin-based conductive epoxy 22
MasterBond MS15 silicon-based conductive epoxy interconnect testing 22
material loss 152
Maxwell Boltzmann distribution 116
membrane-based pressure sensor 48

metal-based additive manufacturing/3D printing 69
metal inductors, liquid 82
metal/ionogels nanocomposite films 102
metallization
  electroless plating 70, 72–74, *73*
  electroplating 70–72
  of ionogels for actuation, energy harvesting and storage 100–104, *101, 103*
  liquid metals 74–76, *75*
  sintering 76–79, *78*
  sputtering and evaporation 79–80
  of 3D printed parts 71–72
metals 69
  liquid 74–76, *75*
metal traces, on complex structures 133
micro-dispensed conductive inks 13
  adhesion evaluation of 13–14, *14*
  vs. aerosol jet conductive inks **20**
  electrical interconnect testing with **16**, 16–18, *17*, **19**
micro-dispensed inks, RF performance of 14–15, *15*
microfluidic biosensors
  design of *107*
  electrochemical 105
microfluidics 74
microplasmas 116
microstrip patch antenna (MPA) 62
MIL STD 883F Method 2002.4 Mechanical Shock 11, 22
multi-functional microsystems 115

nanocomposite resistance 99
NanoParticle Jetting™ (NPJ) 33, 36, 130, 133
National Aeronautics and Space Administration (NASA) 35
natural polymers 104
NextFlex 24
next-generation defense systems 60
nitrate-based ceramics 133
non-conductor 71
non-equilibrium plasma discharges 118
novel emerging materials 96

photo-polymerization of ionogel 102
piezoelectric nanoparticles of barium titanate 103
piezoelectric pressure sensors 48
piezoionic/piezoelectric soft ionogels *103*
planar miniaturized Ka-band antenna 37
plasma-based deposition processes 115
plasma sheath 35
plastronique 117
platinum (Pt) ink 36
polydimethylsiloxane (PDMS) 54, 95, 96, *98*

polyether ether ketone (PEEK)
  sheets 18, 20
  substrates 20, 21
polymer-based additive manufacturing 71
polymer-derived ceramic (PDC) 36, 60–61, *61*
  printed sensors through 60–64, *61–64*
polymers, natural 104
power converters 81
pressure plasmas, atmospheric 116
pressure sensors 47–48, 54, 55
  AM methodologies for fabricating advanced
    59, 59–60, *60*
  fabrication, printing techniques 48–58
  inkjet-printed 57
  membrane-based 48
  piezoelectric 48
  printed 50–51, *52*
  sensing characterization of 62–63, *63*
  solid-state 48
printable conductive inks, formulation of 96
printed electronics
  aspects of 5
  materials evaluation 23
  testing protocols for 12–13
printed PDC sensor 61
  printing setup 61–62, *62*
  sensing characterization of pressure sensor
    62–63, *63*
printed polymer-derived ceramics 60–61, *61*
printed pressure sensors 50–51, *52*
printed sensors, through PDC process 60–64,
    61–64
printers
  Fabrisonic SonicLayer™ 7,200 51, 52
  FDM Titan 51
  IDS Nanojet™ 18–19
  Lulzbot TAZ 4 61
  nScrypt 33, 162
  Plazmod system 118, *119*
  Stratasys FDM Titan 50
  XJET printing unit *130*
printing
  aerosol jet printing 33, 95, 105
  direct write printing technologies 115
  elastomers, 3D printing technologies 82
  inkjet printing 56–57
  for pressure sensors fabrication 48–58
  micro-dispense 5
  polymer-based 3D 69
  3D printing process of ceramic materials
    129–130
pulsed micro-plasma cluster source (PMCS) 95

qualitative adhesion measurements 11

radiofrequency (RF)
  components 32

performance of micro-dispensed inks
    14–15, *15*
radiofrequency (RF) devices
  additive as viable manufacturing method
    161–163, *163*
  extreme customization of 163–165, *164*
rapid prototyping (RP) 47
reflect array antennas 39
resilient hybrid electronics (RHE) 2, 3
resistance 22, 83, 155
  of conductive traces and flexible components
    149
  contact 152
  electrical 76, 83, 105, 108
  of flex conductors 144
  internal 147, 149, 152, 153
  measurements of 23
resistance temperature detectors (RTDs), for 2D
    temperature mapping 125
robotics 93

scratch adhesion test (SAT) tool 5, *7,* 7–8
  high g testing 11–12, *12*
  materials and methods 8–9
  results *10,* 10–11
  testing procedures 9–10
selective laser melting (SLM) 33, 53
selective laser printed pressure sensors 54–56, *55*
selective laser processing parameters 53–54
selective laser sintering (SLS) 33, 53, 54, *54,* 69
sensing characterization of pressure sensor
    62–63, *63*
sensing devices, fabrication of 48
sensor
  electrochemical microfluidic biosensors 105
  fabricated pressure 54, 55
  high-temperature 35–37, *36*
  piezoelectric pressure 48
  soft flexible 93, 96
  soft somatosensory detecting 54
  solid-state pressure 48
Shimadzu AGS-J universal test machine
    (UTM) 63
silica/W-Ni structure 138, *139*
silicoaluminum carbonitride (SiAlCN) 36
silicon oxycarbide (SiOC) 61
silver conductive ink 18, 33
silver inks, direct ink writing of 71
silver-particle-based inks 36
silver-particle-based pastes 36
sintering, metallization 76–79, *78*
size, weight, and power (SWaP-C) requirements 4
slicing algorithm (STL file) 47
slurry-based 3D techniques 61
soft electronics 93
soft flexible sensors 96
soft ionogels, piezoionic/piezoelectric *103*

soft sensors 93
soft somatosensory detecting sensor 54
solid-state pressure sensors 48
space missions, additively manufactured antennas for 37–39, 38
spectrometer 99
sputtering and evaporation, metallization 79–80
stereolithography (SLA) 34, 61
stretchable electrodes 96
subtractive manufacturing 46
supercapacitors (SCs) 104
Supersonic Cluster Beam Deposition (SCBD) 94, 94–97, 99, 102, 105

testing protocols for printed electronics 12–13
tetraethylammonium fluoride (TEAF) 102
thermal considerations, flex and flex-hybrid devices 149–152, 150, 151
thermoplastic elastomer additive manufacturing (TEAM) 51
thermoplastic filaments 71
3D antennas 31, 32
3D copper additive manufacturing (3D-CAM) 39
3D manufacturing 5
of electronics 1–2
3D printed alumina structures 136, 137, 138–139
3D sintered alumina structures 136, 137
3D printed ceramic parts 138
3D-printed electronics 104–108, 107
3D printed inductors 82, 83
3D printed monolithic platforms 105
3D printed NanoParticle Jetting™ ceramic structures 134
3D-printed objects, conductive traces on 104–108, 107
3D printed parts, metallization of 71–72
3D-printed plastics 71
3D printed sintered zirconia structures 136, 137
3D-printed zirconia
aerosol-jetted Molybdenum on 134

Antimony Tin Oxide on 134
electromagnetic test 131–132
3D printed ZrO_2 electromagnetic test 132
3D printing (3DP) 1, 46, 104
advances in 104
of ceramics 61
for printing elastomers 82
process of ceramic materials 129–130
3D Printing Will Rock the World 1
traditional subtractive processing 59
transparent flexible substrates, conductive traces on 96
Tungsten alloy 135

ultra-high-temperature ceramics (UHTCs) 46
ultrasonic additive manufacturing (UAM) 51–52, 52
universal test machine (UTM), Shimadzu AGS-J 63
UV light 61

vacuum 38
deposition approaches 83–84
deposition technologies 80
variable thickness radome (VTR) techniques 35
vat photopolymerization (VPP) 131, 131
vat polymerization 34
Vector Network Analyzer (VNA) 62
VeroWhitePlus 80
very high bond (VHB) tape 57
volatile organic compounds (VOCs) 152

wireless power, inductors and 80–84, 81
wire pull bond testing 18
W-Ni system 136, 137

Yttria-stabilized zirconia, NPJ 36

zinc, deposition of bulk metals 123–125, 124, 125
zirconia 36, 131

For Product Safety Concerns and Information please contact our EU
representative GPSR@taylorandfrancis.com
Taylor & Francis Verlag GmbH, Kaufingerstraße 24, 80331 München, Germany

www.ingramcontent.com/pod-product-compliance
Ingram Content Group UK Ltd.
Pitfield, Milton Keynes, MK11 3LW, UK
UKHW021121180425
457613UK00005B/181